Consequences Of Evolution And Cultural Bias

Consequences Of Evolution And Cultural Bias

Cause and Consequence

Thomas A. Moreau

Copyright © 2015 by Thomas A. Moreau.

Library of Congress Control Number: 2014910803
ISBN: Hardcover   978-1-4990-3794-4
        Softcover   978-1-4990-3793-7
        eBook   978-1-5035-5065-0

All rights reserved. No part of this book may be reproduced or transmitted in any form or by any means, electronic or mechanical, including photocopying, recording, or by any information storage and retrieval system, without permission in writing from the copyright owner.

Any people depicted in stock imagery provided by Thinkstock are models, and such images are being used for illustrative purposes only.
Certain stock imagery © Thinkstock.

Print information available on the last page.

Rev. date: 10/28/2016

To order additional copies of this book, contact:
Xlibris
1-888-795-4274
www.Xlibris.com
Orders@Xlibris.com
543045

TABLE OF CONTENTS

Chapter I	Truth	1
Chapter II	Existence	5
Chapter III	Evolution of Life	15
Chapter IV	Biological Evolution	22
Chapter V	Conscious Awareness and Human Behavior	34
Chapter VI	Evolution of Civilization	40
Chapter VII	Social Morality	44
Chapter VIII	Age of Reason	49
Chapter IX	Judgment	54
Chapter X	Evolution of Cultural Bias Part 1 (Religious Bias)	59 59
Chapter XI	Evolution of Cultural Bias Part 2 (Judaism)	63 63
Chapter XII	Evolution of Cultural Bias Part 3 (Christianity & Islam)	76 76
Chapter XIII	Evolution of Cultural Bias Part 4 (Roman Catholic Church)	91 91
Chapter XIV	Legacy and the Future	102
Chapter XV	Theory of Everything	107
Appendix		113
Epitaph		115
Bibliography		117

ILLUSTRATIONS

1 Paleozoic TL2..26
2 Mesozoic Life Evolution...27
3 Cenozoic Era Time Line..29
4 Life Graphic TimeLine...30

FOREWORD

This Revision is the result of feedback from the original publication. The more technical explanations of the forces of nature has been relegated to appendixes. Explanations of the sequence of events have been clarified, and new information that clarifies events in cultural evolution has been added. Predictions about future consequences remain the same.

    I Look to the future and I see death. I ask myself what should I do with the time that I have left. Who will be effected by what I do and how will their lives be effected. The greatest difference will be to those most influenced by my behavior. What I do will determine the extent that they who are influenced, benefit or suffer. If my behavior is to be judged, then what is the criteria for making such a judgment? The purpose of writing my thoughts the philosophy of living is primarily for self clarification. Most writing regarding this subject matter, are the result of evaluating other "expert" opinions on the subject in order to present a new point of view. My views are the result of my life experiences in order to establish what my philosophical point of view is and the justification for that view. My views are not necessarily new and may have limited academic value. The motivation for writing the book is that a significant portion of educated adults, nearly 50% according to one survey, do not believe in the physiological evolution of the human body. The belief was underscored by the release of a survey of what people believe, taken in North America during the year 2012. The survey was toughly documented and included a cross section of all adults of different ethnic, religious, level of education, and economic backgrounds. Two questions stood out that I found to be particularly troubling. The first was, "Do you believe that the sun goes around the earth, or does the earth go around the sun?". 28% believed that the sun goes around the earth. The second question was, "Do you believe that the physiology of the human body is the result of evolution or an act of creation. 48% believe that it was an act of creation. Similar responses were obtained from the other continents. Every person has a unique philosophy of living that is the result of their life experiences and is as different from every others individual philosophy as their genetic code. Not to be aware of certain fundamental truths, is a failure of the social cultures that denied their members access to these truths.

ACKNOWLEDGMENTS

The author wishes to acknowledge everyone with which I have had life experiences. I am particularly appreciative of those who gave me insight to a contrasting point of view. To name individuals would be exhaustive and require prioritizing. The only one that deserves special mention is my wife, Susan for indulging in the tedious process of editing the script for grammar and punctuation.

GLOSSARY

Awareness	Consciousness.
Dimension	Measurable factor of mathematical expression.
Energy	Ability to change.
Entity	Localized source of energy.
Evolution	Cause and Consequence cycle of change.
Existence	Entity containing energy.
Force	Expression of energy.
<u>God</u>	Creator of the forces of nature that began the evolution of the universe all that exists within. (Biblical definition includes human motivational attributes)
Government	Regulation of social behavior within specified territory.
Inertia	Steady state of physical mass, momentum.
Infinity	Mathematically immeasurable.
Language	Ability to understand and express thought beyond emotional content.
Life	Ability to sustain existence through interactions with the environment via internally generated mobility.
Mass	Self attracting force, having property of inertia, property of matter.
Morality	Standard for social behavior.
Particle	Fundamental entity.

Philosophy	Intellectual beliefs that allow human beings to make decisions contrary to instinct or emotion.
Physical	Occupies space.
Science	Accumulation of data for the purpose of confirming or contradicting a proposed belief.
Space	Multi-dimensional measuring system for locating physical entities and their relative positions.
Time	Dimension that measures change.
Universe	Space containing all physical energy.

CHAPTER I
Truth

Why and How everything that ever happened, or is happening, is by deduction, the bases for establishing the truth. More simply put, if it did not happen, it is not the truth. Reality is the physical consequence of everything that ever happened. What separates Human Beings from all other life forms on the planet earth is that only Human Beings are aware that truth is knowable. The development of language is what made this awareness possible. Language is the mechanical expression of thought. The expression of thought is an accumulative process. Each word in a sentence is a clue as to the intended thought that the sentence intends to express. If I make a statement, then my intention is to duplicate my thought in mind of those who receive the statement. If I ask a question, then my intention is to have the recipient duplicate their thought in my mind. Once a thought has been transferred from one mind to another and committed to memory, their has been a transfer of knowledge. The recipients now knows something that they were previously unaware of, that consisted not only of information, but the source of the information. No duplication is exact because the thought created when the words are expressed, is subject to interpretation by the recipient. The interpretations vary not just as a consequence of each individuals life experiences, but whether the thought was heard or read and the ability of the recipient to understand the language. Failure to understand is failure to convert words into an intended thought process. Misunderstanding is to convert words into an unintended thought process. I know the thoughts that I intended to transfer to anyone who chooses to read this transcript, but I can only hope that these thoughts have significant meaning to those who choose to read these words.

Given these limitations I begin this book with the one absolute and indisputable concept that precludes existence physical or otherwise. Truth is reality, self defining and totally independent of any thought process. Belief of something to be true has no bearing whatsoever on whether or not that something is true. The importance of what any person believes to be true is only significant to the extent that the belief effects one's behavior. If I am able to change my behavior based upon what I believe, then it is in my own self interest to insure that what I believe coincides with the actual truth. Proof that something is true is evidence that supports one's belief in something that, in reality, is true. I may choose to believe that something is true based upon my knowledge about the evidence available to me, but that knowledge is limited. All knowledge is an accumulation of sensory perceptions. I can only know that something is true with relative certainty if the knowledge I poses is difficult for experts to refute where experts are those recognized as having studied the subject matter. If this knowledge was the result of a hallucination or dream, how can one be sure that the source is believable? One may believe that the memory of what was imagined to be proof of one's belief, but only if the perception was not internally generated but rather an externally generated vision.

Knowledge is acquired when the information received is stored, an instant evaluation is made as to the importance and believability of the information received. There are several factors that determine the believability but the most important is the integrity of the source. If you believe that the source has a high degree of integrity, then the prudent course of action would be to trust and verify. Science requires that any proof must be reproducible with access to the tools necessary to duplicate the conditions required for verification of any proposed truth. Until the proof can be provided and verified, any proposal of truth must remain a theory at best or speculation at least. The more data that is accumulated supporting a theory, the more believable the theory.

Science is the study of _how_ the physical universe became the way it is, philosophy is the study _why_ the physical universe became the way it is. If the question is asked; "Why do objects fall to the earth?", a scientist would probably explain the _law of gravity_, but the law does not explain _why_ there is an attractive force between all particles having mass. Science is a methodology for acquiring the data that explains _how_ objects are attracted to earth via the law that predicts their behavior. A person who believes in most religious philosophies, would probably answer the question as the intention of God when the universe was created. Both answers may be true, but only through reason can we determine if both or either can be believed.

All theories began as a speculation of the scientific <u>how</u> or the philosophical <u>why</u>, One's beliefs or personal truths are the result of our understanding the answers to those questions. Determination of those answers require the acquisition and analysis of information. The information acquired needs to be measured in order to determine the degree of believability. The first step in the process of determining believability of any event is to verify that the event actually happened. If you did not participate or observe the event, then trust in the source of information describing the event is imperative. The trustworthiness of the source can be measured most reliably by duplicating the event. If that is not possible then the alternative is to verify the event from other sources. If that is not possible then the last alternative is to measure the reliability and motivation of the source to describe the event. If that is the bases of your belief, then your belief is based on faith. Science is a methodology that requires a more concrete process for verifying an event. If the event actually took place, then their must have been predictable and measurable consequences or changes that took place. Assigning units to the amount of change requires introduces the need for the language of mathematics and is therefore referred to the language of science. Observations, reported observations, or speculations of possible events can then be proposed and <u>how</u> the event occurred can verified as a natural or supernatural event. The result of any such investigation or experiment can only help verify or contradict one's personal belief. If the theory were true then it would have been true regardless of conclusions that were based upon investigations or experiments. In the case of philosophy we are rarely given the opportunity to measure with any degree of accuracy necessary to verify the answer to the question <u>why</u>.

Albert Einstein stated that the truth of all physical phenomenon is relative to a point of reference. This theory can be extended to philosophy by stating that all truth conceived by the mind is relative to the minds point of reference. Therefore, a philosophical proof must substantially rely on pure reason. Reason requires the acceptance of fundamental assumptions that would that would not be acceptable if scientific methodology were applied and has compounded the conflicts particularly between religious philosophy and science.

The single attribute that separates a human being from all other living entities on the planet earth, is the minds awareness of the ability to believe. The most "intelligent" animal can learn to believe that particular actions result in predictable responses, but they are not aware that they have this belief i.e. they have no concept of truth. Seeking truth began with the ability to ask a question during the development of language. All questions begin with the implied thought "I wonder". "I" implies self awareness and "wonder" implies creative imagination.

Each individual seeks their own truth in their own way which may be through science, religion, politics, art, or a wide range of humanitarian endeavors. Regardless of the of the methodology, the reason for seeking truth is to give one's life meaning. Some will explain this need to seek truth as simple curiosity. Curiosity is a problem solving behavior that may require the ability to think logically but does not explain the self motivation to wonder how or why. Many animals other than humans may show signs of curiosity and use logic in solving a problem but only to the extent of solving an immediate problem. The problem may be to seek food or escape from a given situation but never to wonder how or why the situation exists.

As a reasonably educated individual, I accept the laws of nature as defined by the scientific community, as true. I also believe that there is a supernatural form of existence not governed by the laws of nature. Even though science has not yet been able to offer quantitative proof that supports or contradicts a supernatural form of existence, I believe that there is sufficient evidence that a nonphysical form of energy must exist. In order to deny supernatural existence, such a person would have to deny all the evidence that supports such an existence, but also have evidence that supernatural existence does not exist. The foundation of truth is, and will always be in the past. Everything that happens is a continuum of past events exclusive of unconnected spontaneous events. I believe that this is a self evident universal truth and includes the creation of the universe.

CHAPTER II

Existence

(A question of Infinity)

There is a lot of speculation of how the universe came into existence. The most popular current theory is that there are multiple universes that proliferate through a theoretically possible "wormhole" effect. there is <u>no</u> evidence that wormholes exist, but even if they did and were capable of transferring enough matter to create another universe that would not explain the creation of the first universe.

As time moves forward, accumulation of evidence supporting the truth is inevitable. Based on the current level of evidence, logic and reason suggests that at some point in time a supernatural event creating the physical matter happened and that the event had a profound effect on the evolution not only on our universe, but on the evolution of life. Seeking meaning for one's own behavior is to give meaning not only to one's life but to life itself.

Whatever caused the creation of our physical universe, that event began a process of change resulting from the interactions of the forces between the particles created. This is our reality and the reality that must determine our beliefs. Whether this reality was initiated by a supernatural entity or not, is not important. What is important is to understand the forces that brought us to the currant reality so that we can establish a system of beliefs most beneficial to ourselves and mankind in the future. Such forces certainly include the forces of nature. They may also include intervention by a supernatural entity but there is no credible evidence that such interventions took place before the evolution mankind. Even since the evolution of mankind, the evidence is based on speculation and faith in reported events that cannot be confirmed from sources that are substantially biased. Ignorance of the forces that drive evolution, limits one's ability to prepare for the future. The effect is more negative when more people make their decisions based on ignorance or lack of awareness of the short and long term effects that their decisions make on the social environment.

If the existence of anything had no consequence, how can one know of it's existence? The existence is the one requirement that must be established before any thought process can have meaning. Existence is normally defied as physical existence. Isaac Newton defined the basic unit of physical existence being mass. Albert Einstein, in his famous equation $E=mc^2$, defined mass as a source of energy. The expression of physical energy requires the ability to measure motion and the measurement requires both space and time. Even in a supernatural universe, lack of change would reduce the concept of existence meaningless.

Time is the single dimension that transcends natural and supernatural existence in that it is the only dimension that measures change. Time itself does not exist, just as length is the dimension that measures space, but does not exist as an entity. Simply put, anything that ever existed must have existed at a specific time. If we limit such existence to physical existence, then such an entity would have to occupy space. If the universe was created, then there must have been a time that universe did not physically exist. This would require the creation of something from nothing which by definition is a supernatural event. There is significant evidence that this event did take at a specific time and has been labeled the Big Bang. Why this event took place is a matter of speculation. The event was either a random event or an event that was controlled by a supernatural entity. What the nature of the proposed entity was is also a matter of speculation. How the event took place is currently under investigation but the evidence now supports the theory that space was filled with a nonphysical form of energy having an unknown source. This energy field if sufficient in intensity, would result in the precipitation of mass. This was recently confirmed through the analysis of experimental data at CERN laboratories.[1] Regardless of how matter was created during the Big Bang, the event marked the beginning of space and time in this universe. If the event was random then the event must have occurred throughout space resulting in multiple universes. Introducing matter requires one point in space to be uniquely different from all others at the time that the matter was created. The sudden expansion of this matter, resulting in the event being called the Big Bang, introduced the concept of relative motion. Expansion of the space-time continuum into nothingness can only

---

[1] The particle (Higgs Particle) resulted in Nobel prize for physics, 2003.

be imagined as infinite space. There can never be scientific evidence of supernatural existence if such existence cannot be imagined. One can only accept that physical creation happened and the cause of the supernatural event is beyond the ability of the human mind to know or even comprehend with any degree of certainty.

Belief in God as a supernatural entity with a personality defined by various "holy" scriptures, has been indoctrinated into the minds of the majority of human beings. This supernatural entity has been credited with the creation of the universe as an attempt to explain an event that has no other understandable explanation. Although the event was supernatural, there is no way for us to know the characteristics of any supernatural entity that might have orchestrated the event or even if it was the consequence of a "thought" process. To believe that a supernatural entity having the characteristic of a theist "God" was responsible for the creation of the universe is certainly premature, even if such "God" is a universal truth. What happened after the "Big Bang" event through the evolution of mankind would not have changed regardless of "why" or "how" the event took place. Therefore I submit that until the establishment of mankind, any beliefs regarding supernatural events should have no bearing on human behavior.

Thought is the product of physical reactions within the brain just as radiation of energy is the result of physical reactions within the atom, but both result in a non-physical form of energy. The philosopher Rene Descartes established his proof of existence with the statement "I think therefore I am" does not establish physical existence, but only that thought is a form of existence. Science does not recognize any statement as being true unless the statement can be measured and verified. Since thought itself cannot currently be measured with any accuracy any more than gravitational radiation, science can only recognize either as energy sources based on their effect on the surrounding physical environment.

The physical environment is defined as anything that occupies space and has weight (emits energy that attracts other physical entities). This definition requires multidimensional space. There has been an attempt to explain space as single dimensional, two dimensional or three dimensional in an attempt to explain how space made of different dimensions could coexist with observable space. Not only is three dimensional space a minimal requirement in defining physical existence, the general theory of relativity implies that time is also required to define any physical entity as a source of energy. The relationship between space and time required the establishment of the speed of light as a constant.

The speed of light is a ratio between space and time. Declaring that ratio to be fixed simply means that any change in distance requires a proportional change in time. This fixed ratio only applies to light or any electromagnetic radiation through a vacuum (space containing no physical entitles). Any physical entity passing through a vacuum will increase the amount of mass as the physical entity approaches the speed of light. By the time that the speed of light is approached, the physical entity will have maximum mass and maximum distance in a time approaching zero.

Einstein's genius was not only establishing the relationship between space and time, but to show that nothing in the physical universe can be known except as it relates to a point of reference. Scientific understanding is accomplished by proposing a hypothesis and the providing evidence that supports the hypothesis through experimentation. The acceptance of the experimental data establishes the hypothesis as a scientific theory.

The Big Bang is a scientific theory and meets the requirement needed to be established as a valid scientific theory. Since all the evidence now strongly suggests that the Big Bang occurred at a specific time, therefore the universe had a beginning. If that change was a supernatural event, then what change could be measured? The acceptance of God as the answer to an event that cannot be measured, means that the acceptance is based only on faith, primarily on a story written down by Jewish priests in the book of Genesis and declared by those same priests as the word of God. The only change that could be measured was the physical change that began the process known as evolution. The understanding of evolution requires the understanding of physics and the chemistry that defines how the forces that bring about change interact. What follows is an attempt to explain those interactions that resulted in the evolution of the physical universe.

The explanation requires measurement of observable physical changes that can only be done through the language of mathematics. Measuring quantities that are extremely small or extremely large requires the use of the power of ten nomenclature. The first step in measuring is to establish a unit of measurement. In order to measure space, a unit of length is required. The standard unit of length in the community of the sciences is the meter which is equivalent to 39 inches in the more archaic system established by the English. The length of the earths equator would be 40,075,160 meters, a distance difficult to comprehend when compared to one's normal everyday experiences. If one were to measure distances to other objects in the universe, the problem of expressing the distance in meters is greatly compounded. When the distance between stars or galaxies are measured a new unit of measure is necessary to make such distances comprehensible. The unit of measure that has been established is the light year defined as the distance that light would travel in one year. In terms of meters this distance would be about 9.461×10^{15} meters or 9461 with 15 zeros following the decimal point. Consequently, extremely long distances are expressed using the power of 10. Extremely short distances represented as fractions of a meter use the negative sign to indicate that number is a fraction and not a multiple of a meter. The distance that light would travel in a single second would be that fraction represented by a light year in meters (9461000×10^9 divided by the number seconds in a light year (60seconds x 60minutes x 24hours x 365.25 days $=.03155 \times 10^9$ seconds) equals 2998 meters. If one were to measure distances between microscopic organisms, molecules, or atoms much smaller fractions must be expressed using a negative sign when noting the power of 10 thereby indicating how many places the decimal point needs to be moved to the left. The distance of one millimeter would be 1×10^{-3}, the distance of one micrometer would be 1×10^{-6}, and one nanometer would be 1×10^{-9th}. With this understanding of scientific notation we can now proceed with the explanation of evolution as a byproduct of those forces and their chemistry that resulted in the process of evolution.

Energy is the ability to change. All change requires the expenditure of Energy. Beginning with the Big Bang, Energy is being dissipated into space. Eventually all sources of energy will result in the energy of all stars to have dissipated and the Universe will be dark and life-less unless whatever caused the Big Bang is re-initiated. Our physical future can only be understood through the understanding of how Energy is being dissipated. All physical changes require the measurement of time and space. The initiation of such a change requires a source of energy. there are three known sources of the energy that can produce physical change by changing the physical location in space of any physical component. Therefore any physical change must begin with the definition of a physical component whose position will change. Only Matter matters, for without Matter, nothing matters. The unit that occupies space is defined as mass. The very existence of mass is also the first source of energy that can initiate physical change. The change would take place as a consequence of force being applied to a physical component located in space. Mass is both a source and the object of physical change. The force is an attractive force that is emitted through space causing any two or more masses to move through space toward each other. There is no known fundamental unit of mass. If mass were the only source of energy, then space would consist of clouds of mass units that would be continually accumulating and pulling such accumulations apart as a result of the gravitational pull from other mass accumulations.

The second source of energy that can produce physical change is charge. Charge in itself is not a physical component but can attach itself to mass and can change the physical location any mass to which the charge is attached. The force is emitted through space as in the source of mass but charge comes in two forms, one being referred to as positive charge and the other form as negative charge. Like charges repulse each other and opposite charges attract each other. These forces do not effect mass directly but only indirectly via the masses to which they are attached. Unlike mass, there is a fundamental unit of charge that attaches itself to unique units of mass. The unit to which a positive charge becomes attached is defined as a Proton and has a mass of $1.6726231*10^{-27}$ kilograms. The unit to which a negative charge is attached is defined as a Electron and has a mass of $9.1093897*10^{-31}$ kilograms. there is a unit which contains both a positive charge and a negative charge and has a mass of $1.6742286*10^{-27}$ kilogram also defined 1 atomic mass unit (amu). Because mass is a source of energy, any mass represents an amount of energy can be represented in units of energy. The unit of energy often used in particle physics is the electron volt. Using this definition the Neutron, Proton, and Electron having the amounts mass energy; Neutron = 939.56563MeV, Proton = 938.27231MeV, Electron = 0.51099906MeV. The numbers show that an extremely small amount of mass can contain a significant amount of energy. Just a few pounds or kilograms of atomic mass converted to energy, was sufficient to destroy an entire city when the energy was released in a bomb at the end of world war II. The consequence of the interactions between the two expressions of electromagnetic energy was the formation of the first atoms which consisted of a single Proton surrounded by a single Electron in what is defined as an orbital. An orbital is not a single orbit, but rather a multidirectional cloud of orbits around the Proton. If mass and charge were the only sources of energy, then space would consist of clouds of single proton atoms (hydrogen), that would continually be accumulating into dense clumps and then would be pulled apart by surrounding clumps. The evolution of atoms containing more than one proton required a third source of energy.

The third source of energy that can cause physical change has not yet been defined but the effect on the fundamental particles contained in atoms, is well documented. This force originates in the nucleus of atoms and is strongly related to the neutron. The force allows for the creation of a nucleus having more than one proton, separated by neutrons. The neutrons allow multiple positively charged protons to coexist provided that there is at least one neutron for every additional proton beyond the first (hydrogen). The process that allows for the addition of neutrons and protons is called a fusion reaction.

The first fusion reactions took place when the universe was filled with hydrogen atoms immediately following the Big Bang. A small number or reactions took place during the big bang as a result of high energy collisions that allowed neutrons to form and two protons to combine resulting in the formation of Helium Plasma, which combined with other hydrogen atoms resulting in an atom containing two protons, separated by one neutron and surrounded by two electrons. The fusion process released the mass on one electron as electro-magnetic energy with a unique frequency that is defined by the number of protons in the nucleus of the atom. The reaction required a force to compress the three hydrogen atoms to be forced into a single nucleus. This force was gravity condensing huge hydrogen clouds into an entity where the gravitational forces at the center were sufficient to initiate the fusion reactions. These fusion reactions released electro-magnetic energy in the form of light, resulting in the birth of the first Stars. As the population of helium atoms increased, the fusion of various combinations of helium and hydrogen atoms fused into Lithium, Beryllium, and Boron.

Nuclear fusion, primarily of helium atoms, continued to produced atoms having more neutrons and protons in the nucleus. Those having six, seven, and eight protons were produced in robust quantities. These three elements, Carbon, Nitrogen, and Oxygen, accumulated on surface of most planets due to centrifugal force that drove lighter elements away from the center of any planet that spun around an axis and was hot enough to allow flow of the elements that formed the planet. If the Star had sufficient mass density to provide the necessary gravitational force, then the nuclear fusion would continue, producing various quantities of elements having nuclear proton counts from nine to twenty-six (iron). The process known as Fusion Burning (fusion of Carbon with other elements), only takes place in stars that have more than eight times the mass of the Sun, and about 95% of all stars have less than that amount of mass. No amount of gravitational force would allow for nuclear fusion to continue to produce atomic nuclei beyond twenty-six photons. Most stars have been limited to the triple alpha fusion process, where three Helium atoms fuse into Carbon atoms. These stars would evolve into what is called the Asymptotic Giant Branch (AGB) of stars. Furthermore, the energy released from the fusion reactions, beginning and most predominantly the Hydrogen-Helium fusion, of AGB stars would cause the star to expand in size, becoming known as a Red Giant. Once the energy radiated into the surrounding space reduced the mass of the Red Giant necessary to support nuclear fusion, Gravity would cause the star to collapse. The new entity would become an extremely dense entity referred to as a White Dwarf that would continue to radiate visible energy until no visible energy could be detected.

For the stars having more than eight times the mass of the Sun, the collapse of the original Red Giant would collapse beyond a White Dwarf to a entity so dense that the electromagnetic force required to maintain electron orbital's would be overcome and the only particle of such an entity would be the neutron. These entities became known as Neutron Stars. The release of energy during formation of a Neutron Stars are created suddenly during an event known as a Super Nova, and is observed as the star suddenly becomes many thousands of times brighter than the stars not massive enough to have achieved the status of a Super Nova event. The Supernova event would not only release an extraordinary amount of energy, but scattered all the elements produced by nuclear fusion into the surrounding space with such force that collisions between such atoms formed new elements having nuclear proton counts as high as one hundred. Isotopes of Elements having the most number of neutrons over the number of Protons have very long half life's. Carbon Isotope C14 has a half-life of 5730 years, the uranium isotope U238 has a half-life of 4.5 Billion years, about the age of the earth. The space surrounding a Supernova became filled with the material necessary for the formation of both new stars and objects insufficient in size and mass to become Stars. The star referred to as our Sun is just such a star and the planets captured by the sun's gravitational pull, are the result of objects formed by condensation of atoms left over from the supernova that were insufficient in mass to become Stars.

The timeline of the events that took place from the Big Bang to the evolution of our solar system including the plant earth is a story that not only confirms the evolutionary process, but dispels any need for supernatural intervention. The story begins by the determination of how and when our universe's beginning was determined. In 1957 The Soviet Union launched the first man made satellite into orbit around the earth. This spurred the United States into a catch-up mode that established not just the ability to launch a satellite, but justify the need for such an ability. The need for world wide communication was the first commercially viable need that resulted in development of a satellite design strictly for commercial communication by AT&T and Bell Laboratories. The Satellite was launched in 1962 and was used to send commercial (Telephone and Television programs) information to and from Europe. The signals from the satellite had to be received by a sensitive antenna and amplified before sending the information to it's intended destination. The problem was that while amplifying the satellite data, background noise was also amplified. In order to determined the source of the background noise, investigations were initiated. Those investigating the source became aware of a paper written by Jim Peebles, an astrophysicist at the University of Princeton, that the source was radiation left over from an explosion that filled the universe at the beginning of its existence. The characteristics of the background radiation and the prediction that it was from the energy released during the formation of Hydrogen at the end of the Big Bang was confirmed. By calculating the deterioration of amplitude, They could determine that the event must have occurred over 13 Billion years ago. Recent refinements in observable data released by the European Space Agency using data collected from the Plank Cosmology Probe, has concluded that the Universe is 13.8 Billion years old. Because of frequency analysis, they have determined that the source of the radiation was 99% Hydrogen atoms. Even after the evolution of stars and all elements created during that evolution, the visible universe is still 98% Hydrogen.

Because the atoms dispersed by the Supernova would be 98% Hydrogen atoms, the conditions would result in the condensation of a hydrogen cloud into a massive entity with gravity sufficient to initiate the Hydrogen to Helium nuclear fusion cycle of a new star. During the time of new star formations, other atoms will be condensing into smaller entities that will grow into planets that become satellites of the new star. As the smaller entities grew into planets, The gravitational force would cause the temperature of the interior to increase to the point that the planet would become liquid. When these molten planets were captured by the rotational pull of new star, they would rotate in synchronous with rotation of the new star. The centrifugal force of the rotation would cause the elements with less mass to migrate toward the surface of the planet.

The development of highly sensitive instruments and the ability to observe the universe from beyond the earths atmosphere, has confirmed that evolution of second generation stars, planets, and solar systems has been repeated countless times throughout the universe. This describes the sequence of events that resulted in our Solar System and development of the planets including the planet Earth.

CHAPTER III

Evolution of Life

Evolution of life first required the evolution of those conditions conducive to the formation of those molecules necessary for the development of life. For any molecule to form, the elements required must come in contact with each other. The media necessary for elements to achieve the contact that would allow the bonding that defines a molecule is liquid. Liquid to form on the surface of a planet means that the temperature on the surface is sufficient to maintain an abundance of liquid on the surface. Hydrogen and Oxygen would be abundant in the early atmosphere of the planet that is at a distance from the sun that would allow Hydrogen and Oxygen combine forming the molecule H_2O (two atoms of Hydrogen with one atom of Oxygen) known as water, and maintain it's liquid state. Earth not only meets these requirements, but has the other elements soluble in water needed to form the more complex molecules necessary for the evolution of life.

Life is the ability of an entity to <u>initiate</u> a reaction to the environment. The most basic form of life is a single cell or enclosure containing a molecule that is capable of not only duplicating itself, but to generate the cell walls that are sensitive to the surrounding environment. Such a molecule is very complex and is classified as a polymer which simply put is a molecule consisting of smaller molecules chained together. The smaller molecules are made of elements that were available in the huge volumes of water that accumulated on the earth surface immediately after the formation of the earth as a planet circling the star of our solar system defined as the Sun.

During the time that the earth was a molten mass in orbit around the sun, other smaller masses were in the same orbit. Collisions with these objects eventually cleared the orbit of all other objects. As the objects grew, energy released in the form of heat allowed the objects to become molten and form a sphere spinning around an axis parallel with the axis of the Sun. As the sphere cooled, a crust formed on the surface of the planet Earth. The last major object that collided with the earth was the size of the planet mars leaving a huge crater and dislodging a molten mass that was captured by the earths gravity becoming a moon of the earth. The moon acts as a small shield from collisions with other objects such as meteorites and also provides the geology that confirms the earths geology during the formation of the Solar System.

A recent analysis of ancient minerals called zircons suggest that a harsh climate may have scoured and possibly even destroying the surface of the first continents. Zircons, the oldest known material on the earth, offer a window of time back as far as 4.4 billion years ago, when the planet was only 150 million years old. Because these crystals are exceptionally resistant to chemical changes, they have become the gold standard for determining the age of ancient rock according to University of Wisconsin-Madison geologist John Valley. Geologists Takayuki Ushikuba, John Valley, and Noriko Kita published an article in the "Earth and Planetary Science Letters" journal[1] which shows that rocky continents and liquid water existed at least 4.3 billion years ago and were subject to weathering from an acrid climate. According to the publication, no rocks remain from before 4 billion years ago which implies that very high temperatures were the norm. The more recent analysis using an instrument called an Ion Micro-Probe, suggests a different scenario. The Ion Micro-probe measures the isotope ratios of the element Lithium in zircons from the Jack Hills samples found in Western Australia. By comparing these chemical fingerprints to Lithium compositions in zircons from continental crust and primitive rocks similar to the Earths mantle, evidence was found that the young planet already had the beginnings of continents, relatively cool temperatures and liquid water in which the Australian zircons had formed. Simply put, At 4.3 billion years ago the Earth already had habitable conditions. The early earths atmosphere would have contained an extremely high level of Carbon dioxide in addition to Nitrogen and water vapor. Perhaps as much as 10,000 times higher than today's levels. Carbon dioxide naturally combines with water vapor to form Carbonic acid. The combination of an intense greenhouse effect and the acid rain that would precipitate from the atmosphere, would dissolve rocks, even the hardest such as granite. The only ruminants that could be recognized from such ancient conditions would be zircons. Other independent analyses has established the Earth as a separate entity in orbit around Sun radiating energy whose source was the Hydrogen to Helium Atomic Fusion, as having occurred 4.6 billion years ago.

Understanding the evolution of life requires the understanding of chemical properties of the elements and how the elements interact when brought into close proximity at different temperatures and pressures. An excellent introduction or review of how the resulting chemical interactions between the elements produce molecules having their own chemical properties, can be found in the book "The Joy of Chemistry"[2].

In 1859 Charles Darwin published the book "Origin of Species", based on data he collected during a five year voyage on the HMS Beagle that took place twenty plus years earlier. The book was a thorough explanation that attempted to explain how life evolved. The scientific theory was called the theory of natural selection. The reason that he did not publish sooner was that his wife was a strong Roman Catholic and such a theory was in contradiction to Christian doctrine. The power of the church to suppress knowledge cannot be underestimated, even today many people who believe in the bible, do not except the theory of natural selection. The theory was based on subjective evidence which, though overwhelming, did not explain the mechanism that changed the anatomical features of the various life forms. More objective data required the discovery of the DNA molecule. The knowledge of how Deoxyribonucleic acid (DNA), changes from one generation to another and one species to another is still unfolding, but has reached the level that the process of evolution in the development of new life forms is impossible to refute, and the process continues to this day along with the accumulation of data confirming the process.

The reason that the understanding of molecular chemistry is essential, is because every living entity known to exist on the planet earth contains DNA, and how the DNA molecule changes, requires the understanding of molecular chemistry. This molecule is a polymer meaning that it consists of a chain of smaller molecules. All DNA molecule chains contain the same four smaller molecules in different sequences. These four molecules are made from the same elements; Hydrogen, Carbon, Nitrogen, and Oxygen and are identified by the names; Adenine, Thymine, Cytosine, and Guanine or A, T, C, G for short. The chemical description for each of these molecules defines the number of each atom included in the molecule. The Adenine molecular chemical description = $C_5H_5N_5$, Thymine = $C_5H_6N_2O_2$, Cytosine =

$C_4H_5N_3O_1$ and Guanine = $C_4H_5N_5O_1$. The first step in the evolution of life is to show how and when these molecules evolved as a consequence of natural forces being applied to the constituent elements. I have already shown that environment on the earth allowed for the formation of such molecules in substantial numbers. Through laboratory experiments, condition that existed four billion years ago would have resulted in the formation of these molecules. These molecules would have immediately formed chained polymer molecules and become the backbone of DNA. The completion of the DNA polymer would result in the A, T, C, and G molecules in the backbone bonding with complimentary A, T, C, G molecules such that A will bond with T and C will bond with G. The completed DNA polymer now has a parallel chain that is a mirror image of the original backbone, giving the completed DNA polymer the appearance of a ladder that twists and bends to achieve a compact profile.

During the time that DNA polymer was evolving, another equally important polymer was also evolving defined as Ribonucleic acid or RNA for short. The formation of an RNA polymer begins the same way as the DNA backbone is produced using A, C, and G molecules, but instead of the T molecules, a molecule defined as Uracil, U for short having the molecular chemical description, $U=C_4H_4N_2O_2$. RNA has evolved into several variations; messenger RNA (mRNA), transfer RNA (tRNA), and ribosomal RNA (rRNA). Each has a special function in copying a specific sequence of the DNA polymer translating the sequence into amino acid molecules via mRNA, transferring the amino acid molecules to the ribosome via tRNA. The ribosome is a multi-molecular machine composed of two subunits containing one or more rRNA molecules that, in response to input stimuli, produces proteins made from twenty different amino acid molecules. There are twenty amino acid molecules necessary in the construction of proteins. They contain three groups of atoms, the amino group = NH_2, the Carboxyl group = COOH, and the R group which is a side chain that defines the particular amino acid. Proteins are the building blocks of all living structures and provide the ability of these structures to respond to their environment. All living cells are enclosed by a membrane consisting of phospholipids that are two hydrocarbon chains joined to a polar head group that contains a phosphate. The exterior fluid contains molecules that are water soluble. The interior fluid is hydro phobic and referred to as cytoplasm. The interior contains a semi-fluid mixture of small and large molecules including DNA, the variations of RNA and Proteins.

The evolution of DNA, RNA, and Protein took a long period of time even when the environment was ideal. Evidence of when life entities came into existence on the planet earth was established by studying of thin layers of material that show an interaction between micro-organisms and the underlying surface. These earliest known fossils have been geologically dated to 3.5 billion years ago with some analysis have suggested 3.8 billion years ago. All living cells are classified into three types referred to as Domains of life; Bacteria, Archea, and Eukaryotes. Eukaryotes are unique and easily identified because the DNA molecule is separated from all other large molecules in the cytoplasm by it's own membrane. Fossil evidence indicates that the earliest Eukaryote cells date back to around 2 billion years ago and evolved from prokaryote cell that are part of the Archea Domain. Bacteria and Archea Domains used to be classified as a single Domain, but the genetic analysis of their DNA shows that they are genetically separate forms of life.

Regardless of the Domain, every living cell must acquire the energy needed to drive the molecular interactions of evolution. Photosynthesis is a process that converts the energy in light into the chemical energy needed and stores this energy in a molecule called Adenosine-troposphere or ATP for short. Phosphorus is a key ingredient of ATP and can be found in meteorites that plummeted the earth during it's early evolution. The photosynthesis process results in the release of O_2 as a byproduct. Cyanobacteria is an example of an early form of life that used photosynthesis for it's energy. Photosynthesis was not the only source of energy, but it was far more efficient than the others available during the early evolution of life. Not only did it allow for rapid growth of biomass, but also allowed for oxygenation of the oceans during a time when all the available oxygen was bonded with Hydrogen to produce water, but also bonded Carbon to produce Carbon Dioxide, or bonded with phosphates in the molecular construction of ATP. The release of free oxygen initially bonded with minerals that resulted in precipitation of iron oxide into large iron deposits. Free oxygen available to the continuing evolution did not begin to measurably increase before 2.2 billion years ago.

Geologically, the events that describe the evolution of the three Domains of life took place during a time period known as Archeon Eon and covers the time period from 4.0 to 2.5 billion years ago. By the end of this Eon, Continents had emerged, cyclic re-melting and reformation of rock through lava flow (igneous differentiation), allow the lighter minerals to float over the liquid mantel. Hot spots resulted in volcanoes both on the continents and under the oceans that often produced islands. Eons are divided into periods of time referred to as Era's. The Paleoarchean Era covered the time period from 3.6 to 3.2 billion years ago. There is no specific rock layer that separate the Era from previous or later levels, instead the Era has been defined chronometrically by determining the time required for sequential events to occur. Both bacteria and archea evolved during this Era, which implies that the "Last Universal Common Ancestor (LUCA), Existed at the beginning of this Era. The first organisms were likely non-photosynthetic, using methane and ammonia for their energy needs. All life from this Era was anaerobic meaning that the organism does not require oxygen for growth.

Free oxygen is toxic to anaerobic organisms and the rise in free oxygen would have destroyed most of the earths anaerobic inhabitants. From their perspective it was a catastrophe and marked the first mass extinction event in earths history. This event would not have occurred until oxygen had risen to account for 20% of the earths atmosphere. This would not occur until the end of the next Eon, beginning 2.5 billion years ago. Long before oxygen reached 20% of the earths atmosphere, free oxygen reacted with the atmospheric methane, reducing the concentration and it's "greenhouse" effect, thereby triggering glaciations. Beginning as early as 2.3 billion years ago, glaciations gradually covered the continents and froze the oceans, becoming known as a snowball earth event which lasted for 300 to 400 million years. Variations in energy coming from the sun and volcanic activity eventually resulted in a buildup of CO_2 resulted in rising temperatures and the end of the snowball event. Oxygen levels continued to rise initiating other snowball events. One at the end of the Proterozoic Eon that played a roll in the first recognized mass extinction of the earth and marked significant changes in the evolution of life.

The study of evolution requires classifying the sequence of events that allowed for future evolutionary changes. This requires a clock that can be be used to determine the time at which events occurred. By taking core samples of material deposited on the surface of the earth over extended periods of time, we can analyze the molecular composition of ice and minerals from the distant past. The Atoms that made those molecules can be identified by the number of protons in the nucleus. We also know that every element contains a unique number of Protons plus a number of Neutrons equal to or greater than the unique number of Protons. If the element is Carbon, by definition, the number of Protons is six and the number of Neutrons is at least six, but may be more. The number that equals the number of protons plus the number of neutrons is referred to as an Isotope. Carbon 12, Carbon 13, Carbon 14 are all Carbon Isotopes. All elements having an excessive number of neutrons will be unstable and loose those excessive neutrons over time. The rate that they loose those excessive electrons can be mathematically calculated such that if a sample contains a percentage of an isotope having excessive neutrons, half of that isotope will be degraded in a specific period of time. This period of time is referred to as half-life of the elements isotope. In the case of Carbon, the half-life of Carbon 14 is 5730 years plus or minus 40years. Heavier Isotopes have longer half-life's with uranium 235 having a half life of 703.8 million years. The accuracy depends on the quality of the sample and therefore limits carbon dating to typically plus or minus 100 years, but still allows us to determine the time at which molecules were formed from the formation of a crust on the earths surface to the present.

CHAPTER IV

Biological Evolution

From the reference points that have been established, a timeline can be constructed that defies the sequence of events describing the evolution of life. The graphic timeline on page 26, confirms the events that occurred on the planet Earth, and would presumably occur on any planet having the same elements available in liquid water. The roll of mass extinctions that provided the pathway for multi-cell evolution clearly shows life-forms unique ability to adapt to changes in the environment.

The ability of multi-cell life-forms to adapt much more quickly reduced the time span from over three billion years of single cell evolution to less than 700 million years of multi-cell evolution bringing us to the evolution of primates. The geological time at which major environmental changes took place has been classified into Eons which are divided into Eras, that are in turn divided into periods. Most of the single cell evolution took place in the Archean and Proterozoic Eons over the last three plus billion years. Evolution of multi-cell life began at the end of the Proterozoic Eon during a time period classified as the Neoproterozoic Era that began one billion years ago and more specifically during the Ediacarian Period of the Neoproterozoic Era, that began 635 million years ago.

What part Life itself played in the mass extinctions is continuously being investigated and whether the evolutionary process would take a similar course on other planets may not be known for centuries but some form of evolution has certainly taken place on the billions of planets we now know that exist in the visible universe.

Once Free Oxygen became available in the atmosphere and in water, a new evolutionary process began. Oxygen availability allowed for the evolution of multi-cell life forms and the evolution of circulatory systems necessary to allow for the metabolism of all cells while allowing them to perform specialized functions. All fossil evidence of multi-cell life, where cells specialized into symbiotic relationships with other groups of cells, begin during a period that has been defined Neoproterozoic Era of the

Proterozoic Eon, (1 billion to 541 million years ago). More specifically, the last two periods of the Neoproterozoic Era, the Cryogenian (850-635 million years ago) and Ediacaran (635-541 million years ago) Periods. During these two Periods of time their was a dramatic increase in the oxygen level in the atmosphere rising to 28% at the highest point. This rise triggered several glaciations that lead to "snowball" earth events, each separated by a warm cycle. The first snowball-warm cycle took place during the Cryogenian Period which placed stress on all life forms. Those life forms that survived the cold were able to evolve into forms that could take advantage the resources available in the new environment. During the Cryogenian Period the evolution in life forms resulted in the first multi-cell organisms, the first plant life, the first animal life, and indications of the first sexual reproduction. These evolutionary steps all originated from the Eukaryote Domain of living single cell organisms. Algae became the first example of multi-cell plant life, and is responsible for an ever increasing rate of free-oxygen production. Sponges are the earliest example of animals.

All life during the Ediacaran Period was soft bodied, there were no shells, bones, or teeth. Because of the high level of oxygen, nearly all anaerobic organisms went extinct. The end of the Ediacaran Period was marked by the last snowball-warm cycle and the first Mass Extinction event. Geologically, The super-continent Rodinia, began to split apart 900 million years ago, a time when only single cell organisms exited and only in the oceans. Over the next 600 million years, Rodinia split apart and converged into in new super-continent called Gondwana which broke-up reconverging into the super-continent Pangaea. Beginning about 250 million years ago Pangaea began to break-up, into a north hemisphere continent called Laurasia and a southern continent called Gondwanaland, which in turn broke-up into the continents that exist today. Evidence of continental movements just described, was established by comparing rock formations and the radioactive dating of their radioactive isotopes. Beginning about 300 million years ago, multi-cell life began to migrate from the oceans to land and their fossils also confirmed continental drift.

Every living entity has a genetic identity which defines all of the physical characteristics of the identity. In addition, the generic code defines how the entity will, at least initially, respond to it's environment. It is logical that if the response is successful that the species will pass this genetic information on during reproduction. This process explains the survival of a species provided that the environment does not change drastically in a short period of time, which would compromise the ability of the species to successfully respond. This process does not explain evolution from one species to another. A new species requires a new genetic code, one that allows for the new species to survive in an altered environment. One method that allows for offspring to have a genetic code containing genetic information from two separate genetic codes and is defined as meiosis, commonly referred to as sex. Meiosis is a special type of cell division that produces germ cells (eggs and sperm). Through the act of sex, the sperm cell combines with the egg to produce a unique genetic code which results in a large variety of genetic codes within a species but does explain the change in genetic code that would result in a new species. There is another change that takes place during an entities lifetime that are reflected in the germ cells of each individual parent. These changes may appear to be insignificant, but over several generations, real changes in both physical characteristics and behavior become obvious. Evidence of this process is most clearly demonstrated in dog breeding. The process can result in a new species as Darwin demonstrated without the advantage of knowledge about the genetic mechanism that made such changes possible.

By the beginning of the Paleozoic Era, meiosis had become the preferred method of reproduction for Plants and almost the exclusive method for Animals. The consequence was an explosion of life forms in the Paleozoic Era. Early Mammals were established throughout the super-continent Pangaea before it began to break apart and are represented in all the continents if not living, as fossils. By the middle of the Paleozoic Era, Laurasia had split into North America and Eurasia with a narrow land bridge between northeastern Siberia and northwestern North America. Gondwana land had divided into the Antarctica-Australia to the far south and the South America-Africa to the north. India had separated from the Antarctica-Australia land mass. Mammal evolution in far south provided an isolated environment that allowed Marsupial, to flourish and Monotremes to coexist. Monotremes lay eggs into a pouch, Marsupials give birth into a pouch, where they are nourished for a period of time before they are able to live separate from the mother. By the end of the Paleozoic Era, South America and

Africa had separated with Africa having drifted north into Eurasia and Australia had separated from Antarctica.

The evolution of multi-cellular life and the geological events that had significant effect on that evolution, is shown on the following graphics. The first graphic shows a table of major events that lead to multi-cellular life during the Paleozoic Era, 541 to 250 million years ago (Mya), and Periods of that Era, followed by a table of major events during the Mesozoic Era, 250 to 65 million years ago. Mass Extinctions account for sudden changes in the rate of change, whether those changes would have taken place at a slower rate is not knowable.

524Mya Growth in multicellular life almost exclusively in the Eukaryote Domain, requiring the Domain to be subdivided according to Type of life forms. Kingdoms, Phylum, Class, Order, Family, Genus, and Secies were all established during this Era.

PALEOZOIC ERA					
Cambrian	Orovician	Silurian	Devorian	Carboniferous	Permian
488 Mya	444 Mya	416 Mya	359 Mya	299 Mya	251 Mya

Cambrian: Explosian of mult-cell Life and classifications of life forms, first Virtebrates and exoskelitons. Athropods such as trilobites were Vertibrates with exoskeltons that latter evoved into insects, spiders, and crustations such as lobsters.

Ordovician: All multi-cellular life existed in the oceans, the first multi-cellular Mass extinction took place toward the end of the period that caused more than than 50% of living Families to become extinct.

Silurean: Melting glacial formations caused significant rise in sea levels And the apperance of coral reefs. The rapid spread of jawless fish and the first appearance of jawed fish as well as fresh Water fish.

Devorian: The Period is marked by one significant and a few lesser Mass Extinctions that caused 19% of Families and 50% of Genera of sea Life to become extinct. Jawed vertibrates and Land life seem to have Not been effected although land life was mostly early forms of plant life and Insects. All fish cassificatios date to the Devonian Period.

Carboniferous: Acient tree-like plantsgrew abundantly in low lying, swamp-like Environmentsbecame the source coal and oil deposits.

Permian: The greatest Mass Extinction that has ever occurred in the history of Life on earth ended the Paleoic Era as well as the Permian Period. Over 95% of marine Species and 70% of terrestrial vertibrate species Became extinct. 57% of all Families and 83% of all Genera were Gone. Pangea, the last Super-Contient was fully formed.

Timeline and events that marked the evolution of multi-cellur life

248Mya : Evolution of Terrestrial Life from amphibians to mammals, Age of Dinosaurs

MESOZOIC ERA		
Triassic	Jurassic	Cretatious
▲248 MYa	▲206 MYa	▲144 MYa ▲65 MYA

Triassic — As with all mass-extinctions, the Period following was a time that the surviving life forms expanded rapidly in numbers and variations. The two Kingdoms that had the greatest expantions were the Plant and Animal Kingdoms. Conifers, Palms, and Ferns covered the landscape. Insects that feed on the plants and animals that feed on insects soon followed. Anphibians emerged from the sea. Turtles Reptiles and the first flying vertibrates immerged.

Jurassic: Single cell animals proliferated in the oceans, the first bony fish ememerged. Dinosar species both herbivoirs and carnivors were prolific and grew to enormas sizes. Mammals continued to diversify, though remained Small and had little impact.

Cretacious: Pangea had seprated into Laurasia in the northern hemisphere and Gondwanaland in the southern hemisphere, Fossil evidence shows that terrestrial evolution took different paths on the seperated Continents. The period ended with another mass-Extinction event that brought an end to the Dinisaurs And the Mesozoic Era.

The evolution of life that took place during the Paleozoic Era clarifies two properties of the evolutionary process. First, the rate of change that takes place is constantly changing. There are periods where very little change takes place, periods where huge changes take place, and periods that appear to reverse previous changes. Second, with the consequences of change is a new environment that provides opportunities for new changes. Evolution is a process that began with the forces that initiated the process and will continue so long as those forces are present. The evolution of multi-cell life forms is a confirmation of how the process inevitably results in a more complex environment. The Era that follows the Paleozoic Era continues to confirm the process and will do so with or without supernatural intervention.

The Mesozoic Era demonstrates how life forms are dependent on the environment The second table shows the timeline of the Mesozoic Era, 250 to 65 Mya, that emphasizes the significant events that took place in the three Periods of the Mesozoic Era. The evolution that took place during the Mesozoic Era, 252 Mya to 65 Mya, confirms the difficulty in predicting what changes will take place or when significant events will take place.

A Time Line of events that took place beginning with the Ediacarian Period of the Neoproterozoic Era through the Paleozoic Era, and the Epochs that lead to evolution of Homo-Erectus, covers the events following the last major mass extinction that allowed mammals to evolve into an animal capable of understanding the concept of truth and laid the foundation for the evolution of modern man. The Time Line is preceded by a table of major events that took place during the Cenozoic Era. The last timeline shows the time line of the Cenozoic Era, (65 to less than 1 Mya) the periods and six of the seven Epochs, that identifies significant events as they evolved during those Epochs.

CENOZOIC ERA					
Paleogene			Neogene		
PALEOCENE	EOCENE	OLIGOCENE	MIOCENE	PLIOCENE	PLEISTOCENE

▲ 65 Mya   ▲ 55.8 Mya   ▲ 33.9 Mya   ▲ 23.0 Mya   5.3 Mya   2.6 MYa   0.1 MYa

EPOCH

Paleocene: Following the mass extinction that eliminated the Dinosaurs, the climate was tropical and the continent of Laurasia began to split into North America and Eurasia. Gondwanaland also began to split. Rodent size mammals grew in number and diversity, some retuning to the oceans, including the first whales.

Eocene: First Primates or mammals with flexible hands having Apposing thumbs, bifocal vision and larger brain to body weight ratios. Hoofed animals, bats, many species Birds originated in this Epoch, living In Jungles and rain Forests that began to change as the atmosphere began To cool giving way to deciduous forests.

Oligocene:: Grasses that resulted in big changes allowing Herbivores to grow in size and diversity along with carnivores that feed on them. First primates appear in Africa Antarctica covered South Pole and separated from Australia establishing ocean currents that caused glaciation to to grow across the continent.

Miocene: Continental Tectonics separated them into all the continents that exist today and produced the major mountain ranges. Kelp forests In oceans emerged and grassland greatly expanded. Ice-caps formed at the poles and seasons became established.

Pliocene: Last Epoch of the Neogene Period, The first Hominids appear in Africa

Pleistocene: Glaciation cycles in the northern hemisphere continue to the present time an is the most significant factor that effects sea level and climate change. Homo Sapiens,only surviving member of Genus Homo, populated Africa and all of Eurasia.

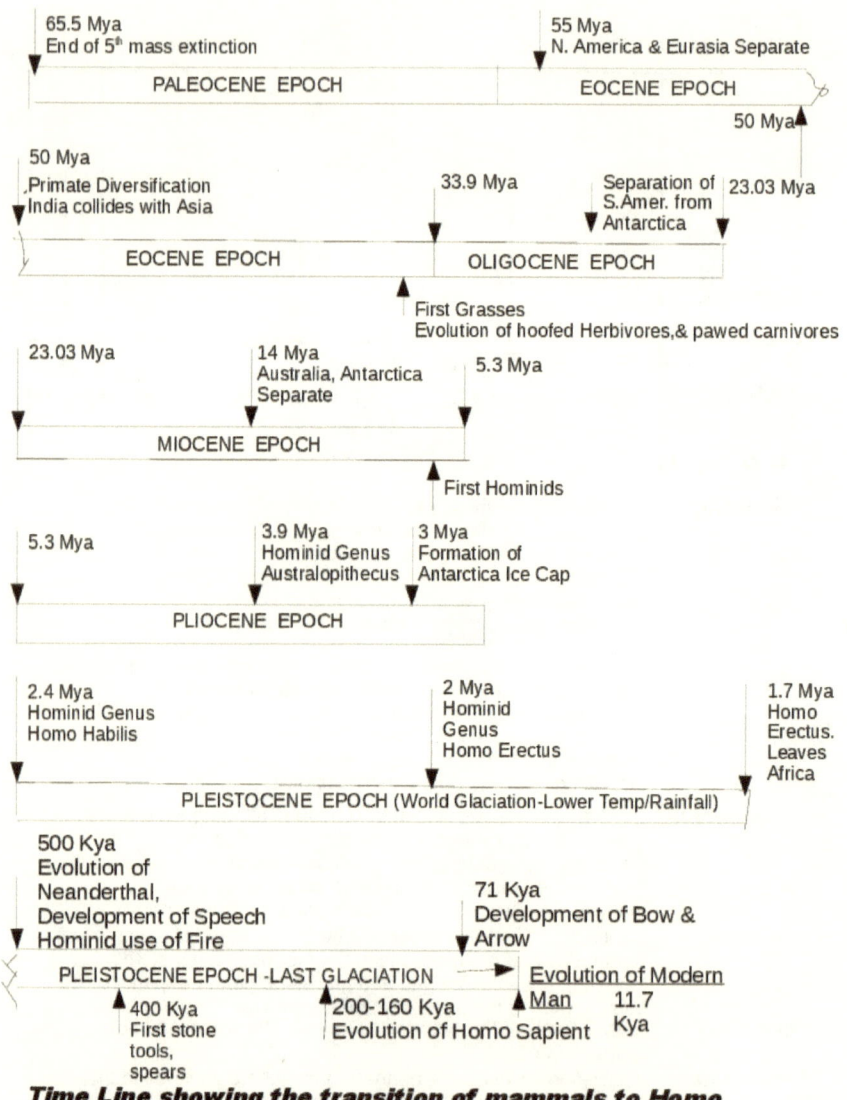

Time Line showing the transition of mammals to Homo Sapiens following the last major mass extinction.

The continents continued to separate during the Cenozoic Era, Africa reached the Arabian peninsula, India collided into Asia creating the Himalaya mountain range. Australia separated from Antarctica and other major mountain ranges formed as a consequence of plate tectonics. All of these factors produced a variety climates unique not only to the individual continents, but also unique climates within the separate continents. The expansion of terrestrial life-forms following the mass extinction that drove the Dinosaurs into extinction had to initially take place in the Plant Kingdom of life-forms because all life-forms in the Animal Kingdom depend on the Plant Kingdom for their nourishment either directly ort indirectly. The Plant life forms that quickly recovered and dispersed over large areas were the tree-like plants which spread seeds forming dense jungles.

The Cenozoic Era covers the periods that identify the changes that took place between 65 million years ago and the dawn of civilization. The evolution of life forms that took place were in response to the unique environments found in the different regions of the different continents. The most significant events in regard to the evolution of mankind took place in south Africa, the birth place of Homo Sapiens.

The climate in south Africa was one of the ideal for the formation of Jungles and produced large quantities of seeds, nuts, and fruit that was needed for animal evolution to take place. In order to access this prodigious source of energy, animals needed to adapt their body plan that would allow them the ability to reach this source of food. Although the large Dinosaurs had died out, one sub-Family of Dinosaurs had evolved with feathers, probably for keeping the body warmer, developed those feathers into the ability to fly. Not only did this allow them to escape predators, but gave them access to the bountiful food source in the trees. The other sub-Family of animals that developed the ability to access this food source, were the early primates. They learned to climb trees and used their bifocal vision along with their hand-like appendages to move through the dense jungle, Those that acquired this ability were given their own classification of life forms, Haplorhini Catarrhini which is a Primate sub-Order, later evolved into the Families of Tarsiers, Monkeys, and Apes. Haplorhini diverged from the other Primate Sub-Order 63million years ago. that evolved into the family of Lemurs. All of these evolutionary events took place During the Cenozoic Era.

The last two epochs cover the evolution of Hominines, a taxonomical tribe of the subfamily Homininae that is comprised of three sub-tribes: Hominina, with one Genus *Homo; Australopithecina*, having several extinct genera, and Panina, with one genus Pan, (Chimpanzees). Hominini is new sub-Family that includes four Genus, Australopithecus, two variations of Australopithecus, and Homo. The one physiological characteristic that identifies the Hominini Sub-Family is that they are all bipedal. When the climate changes caused forests to become less dense with many containing conifers, the primates needed to descend from the trees to move through the forest and seek other sources of food. Grasses evolved into savannahs that provided for the expansion of herbivores. The need to observe their environment required the Hominines to stand upright. The new food source was insects and small rodents, making them omnivores.

The Genus Homo developed the ability to make tools for cutting that allowed them to make Clubs by attaching a stone to the end of a stick and Axes, both of which allowed them to protect themselves and hunt larger herbivores. The first fossil evidence that represents the Genus Homo has been defined as the Species Homo-habilis because of it's ability to use it's hands for making tools. Access to larger quantities of meat provided the protein necessary for the physiological growth of the body and the brain. Homo-habilis has become a marker for the beginning of the Pleistocene Epoch, 2.8 million years Ago. Homo-habilis was soon replaced by Homo-erectus having a larger brain and body and the ability to make more complex tools and weapons for hunting larger prey. Fossil evidence dates Homo erectus to about 2 million years ago. Homo-erectus also acquired the ability to start and control fire. Fire was used for defense, heat during seasonal low temperatures and cooking food, primarily meat. Cooking meat made the protein significantly more digestible and enhanced the brains ability to make more sophisticated tools and weapons that resulted in spears. The increasing dependence on hunting herbivores of the savannah, and the seasonal climate changes required Homo-erectus to migrate with the herbivores as they migrated to find food sources replacing the savannahs that died off when the climate cooled each season. Homo-erectus migrated out of Africa across the middle east. Fossil evidence show that Homo-erectus reached Indonesia, China, and north Asia. Although Homo-erectus eventually became extinct with the last fossil evidence being around 150, 000 years ago, several replacement species evolved along the way. According to fossil evidence currently available, the most significant replacement was Homo-heidelbergensis. This species apparently also evolved in Africa less than one million years ago and migrated north into west Asia and Europe. Another migration followed the eastern path and evolved into a new species, Homo-denisovan. The northern migration evolved or were replaced by the species, Homo-neanderthalensis. Both species have fossil evidence indicating their origin to be around 500,000 years ago. The most significant species to have evolved from Homo-erectus was Homo-sapient sometime between 150,000 and 200,000 years ago. The differentiation of these three species that derived from Homo-erectus, were determined not just by the physiology of their fossils, but by the analysis of their DNA. The DNA evidence indicates that at least between Homo-neanderthalensis and Homo-sapient there was interbreeding. Homo-sapient is the only species of the Genus Homo still in existence with the others having gone extinct by 40,000 years ago. How Homo-sapient was able to survive all the climatic changes that have taken place over the last 200,000 years is still being debated, but the development of the brain, not in size but in how it is wired, is agreed to be the fundamental cause.

The next chapter will be concerned with understanding of how the brain became rewired as a consequence of the evolutionary process.

CHAPTER V

Conscious Awareness and Human Behavior

As the evolutionary process proceeded, animal life adapted the ability to respond emotionally to the environment. Although the sex drive is a biological phenomenon, an emotional response to this drive is responsible for the brain's action. That makes emotion a fundamental force behind animal behavior. Emotion can be measured on a scale that rates emotion from fear to desire just as temperature is on a scale that is rated from cold to hot. Circumstances and memory of past events will modify the brain's response sensory input that triggered the emotion, but the sensory input is the trigger that resulted in the animal's behavior. Instinctive behavior does not require the recall of past events, instead the response is based on "genetic memory". An animal is instinctively motivated to promote the survival of the genes. Any behavior that is contrary to the survival of the genes is indicative of either a genetic disorder, or a conscious decision to over-ride instinct. Every living entity is born with instinctive behavior designed to insure it's immediate survival. Other instinctive behaviors become active as a result of growth but are instinctive because nothing has to be learned in order to invoke such behavior.

Emotion is equally important in explaining behavior of any living entity having a central nervous system with the ability to remember past events. The entity is then able to compare sensed information with remembered information and behave in a manner most advantages to the entity. There is an emotional response to every thought. The process requires establishing priority. Genetics provides the ability to make decisions, the environment provides the need. The process of making decisions includes assigning values to the information received from the environment according to it's relevancy. Determining value and relevancy is an extremely complex process, especially when each value has both an instinctive and emotional component.

Human behavior adds another step to the process, the ability to reason. Regardless of the decisions made reason takes into account one's knowledge of their own identity. Beginning with the awareness of one's own identity humans are able to control their own behavior. This ability results in the desire to control the behavior of others. There are three components that motivate human behavior; instinct, emotion, and reason. The source of instinct is genetics that emphasizes not only self survival, but also the survival of those that whose genetics are most closely related. This often requires putting the interest of others before one's own interest. The application of emotion and reason can override those interests.

The process that differentiates between instinctive, emotional, and reasoned behavior is the resulting associations that the mind makes with genetic memory and memory of past events. There is no knowledge on which to base any judgment that had not been acquired through the senses. Even one's own existence can only be deduced through the information acquired via the senses. When the philosopher Descartes declared "I think therefore I am." can only be true if there are memories about which to think. Those memories are a reflection of sensory provided information and may be distorted but are the only source of reality that is available to the mind. Any suggestion that the mind can function independent of memory, is to ignore the reality upon which such thoughts are based. Personality of an individual is the consequence of long term memory. Observation of a person suffering from Alzheimer's demonstrates that lose of memory is synonymous with loss of personality.

There are five undisputed senses; touch, taste, smell, hearing, and sight. There is at least one additional sense that I refer to as the Instinctive sense or genetic memory. The reason that instinct is not usually considered a memory, that the recall of such memories is not available to the conscious mind, rather only to the sub-conscious mind. Continued research into the mechanisms of memory will soon clarify that instinct is indeed a memory that can be invoked as a sixth sense. This is a sense that is often associated with religious beliefs. Memories associated with dreams or hallucinations appear to have an unknown source often presumed to be an external spiritual entity. Only if the memory has a predictive element or describes an event that took place somewhere else can the memory be validated as being from an external source. The verification that the event did or will occur at a specific time and place that violates the laws of probability. In biblical terms, those having such memories, are referred to as prophets, the accuracy of the reported "prophecy" and fulfillment will always be open to question. Those who believe in the source of such predictions require an element of faith that can never confirmed or disproved through scientific analysis.

What makes human beings unique among all living entities is the way we think. There are anatomical differences that may be required in order for us to think the way we do, but the thinking process itself is what makes us unique. We determine how human beings think by observing behavior. Human life can be most accurately defined by defining the behavior that is unique to and results from the intelligent thought process beginning with the ability to think logically. The bases of logic is the If-Then statement. Computer programs are based on the ability to make decisions based on the logical statements If (condition)-Then (go to). Many animals use logic to solve problems, a squirrel trying to access seeds intended for birds is a common example. Logic alone does not provide the ability to reason.

Self-awareness is a step in the evolutionary process that allows for the possibility of reasoned thought. Three biological developments must take place before self-awareness can be achieved. First, the brain must have the ability to remember life experiences. Second, the brain must have the ability at compare different memories. Third, the brain must be able to create a self image from these memories that defines the individual as a separate and unique entity. Humans are not the only animals to achieve self-awareness. Recent studies have shown that elephants, dolphins, and chimpanzees are self-aware. Chimpanzees have demonstrated that their ability has been extrapolated to see their trainers as individuals that could be manipulated to provide them with special treatment.

The thought process that truly makes human beings unique is creative thought. The process of being creative is what allows us to make decisions not just based on genetics, or responses to life experiences, but also responses to our imagination. The creative process can be defined as the expression of "free will". Some would suggest that evolution results in a fatalistic future and denies free will. The truth is that "free will" is the consequence of a human thought process. The origin of this thought process could be the result of a super-natural intervention at a specific time, or the natural continuation of the evolutionary process.

Philosophically, the question of how life began is important to the extent that if outside intervention were required, then knowing the nature of this intervention would give us a clue as to it's purpose. The most common belief is that nature of the supernatural intervention is defined as God. God is a name that is given many attributes, but the "assumption" is that God created life on the planet earth with the express purpose of establishing a relationship with humans. This evokes the question, "Why create such a vast universe with billions of galaxies, each having the ability of establishing life on potentially millions of planets?" A far more probable point in time that would justify the intervention of a God-like supernatural intervention would be after the evolution of Homo-sapient. Until humans are capable of moral behavior, why provide humans with a soul.

Whether such intervention took place or not, changes the way the brain processes information was required and resulted in a new level of awareness not just based on what one experienced, but awareness of that which one has not experienced. The most obvious way of obtaining information that one did not experience, is through use of language. Language allowed the transfer of experience from one mind to another. Language is not just communication, All life forms can communicate in one way or another, but only communicate their instinctive, or emotional status.

The evolution of language was a complex evolution of how the human brain was wired that began during evolution of Homo-erectus and continuing through the evolution of Homo-sapient. The evidence that is convincing to any student of anthropology that humans had achieved "free will" is when humans transitioned from hunter-gathers to herders- farmers. The changes that were needed to develop a language involved the coordinated control of the Larynx, Ears, Memory, and conscious interpretation of specific sound sequences. Birds have the ability to produce the entire range of sounds that can be produced by the Larynx, hear the entire range, and remember entire sound sequences. They can also mimic these sound sequences but do not have the ability to assign meaning to specific sound sequences.

Assigning meaning to specific sound sequences raised the level of communication between humans that allowed for a social structure unachievable by any non-human life form. True language began with the use of sound sequences that we define as words and the ability to use those words to ask questions. Every question begins with the premise "I wonder". The consequence of such social interactions were the establishment of rule of behavior within the social group.

The immediate need of the body is to breath. This need does not require conscious thought unless the air is polluted smoke or there is some other obstruction restricting oxygen from reaching the lungs. In such a situation, the mind will invoke conscious thought and the action necessary to resolve the problem. The mind also has the ability to override the needs of the body. Examples would be to save another persons life, martyrdom, or suicide. These are a few examples were the psychological needs exceed the needs of the body. The decision process would be based on a conscious evaluation of self worth. If one chooses not to risk your life, then you have determined your self worth to be greater than the consequences of risking or deliberately ending your life. How the mind makes such decisions is based on a personal moral standard that developed from the time that the mind became aware of it's ability to override the needs of the body. These beliefs are strongly influenced by the instinctive motivation of survival of one's genetic identity as an individual and a genetically related group.

Morality is a learned behavior and an extension of logical problem solving behavior. Learned behavior reflects those beliefs that the individual is obliged to society in which one is a member. Morality defines behavior as being right or wrong to the extent that it reinforces one's personal belief or truth. Immoral behavior hides what one believes to be the truth and results in a feeling of guilt. A pattern of moral behavior can only be changed by awareness that a belief was not true or of a revelation of contradicting truth. Such events will always introduce a degree of trauma. The more traumatic, the greater the change in behavior. If a person were on a deserted Island, what behavior would be immoral? Only behavior that would give rise to guilt would be considered immoral. Such feelings would be the result of learned standards of morality and if one had not been taught such standards, immoral behavior would not be possible. The logical conclusion is that moral behavior is intended to benefit society and not the individual.

Social behavior determined by moral law encourages behavior by the individual that is most beneficial to the social group. This behavior in turn, encourages specialization in those behaviors that the individual is most adept. The specialization not only allows the social group to grow, but greatly increases the variety of life experiences to which members of the social group are exposed. The greater the number of life experiences, the more likely that an individual will make an association of such experiences and create a thought image that is truly original. Thus began the creative thought process, increasing the rate of specialization and growth of the society. The rate of growth can also be increased by the availability of natural resources, but morally established rules by which the members of the social group need to conform, are mandatory in order for the individual to survive within the society.

As the society grows, government is established to define and enforce the rules of morality. Because each individual interprets the moral standards of the society, according to their own genetic heritage and life experiences, government will most closely reflect the moral standards of those who establish the laws that enforce those standards.

Any society will exist, barring military intervention, so long as the government enforces the moral standards that reflect the will of the majority. The government may try to change those moral standards by changing the rules or the rules that the government enforces, but if those rules do not benefit the majority, even the most authoritarian government will fail.

The first person to seek the truth may have sought the truth about making a better tool. Whether successful or not, that person gave life purpose. To the extent that people seek truth about what is important in their lives, they give their lives meaning. Not all humans seek the truth and nobody seeks the truth about anything until they reach the age of reason. Acquisition of truth is power and the advantage is control over those who do not poses or are unaware of the truth. The desire to keep the advantage leads to secrets that result in lies an corruption.

If one is judged by that persons ability to make the right choices that is based on their own standard of morality, then the individual should seek to understand how that standard of morality was obtained for their own benefit. The understanding one's individual morality begins with the genetically driven motivation of survival followed by the social indoctrination that results in the moral standards of the individual. The society that is successful is deemed to be civilized. The first societies that have been designated as civilized are those that transitioned first into herders and then into herder-farmers. Evidence of such societies date to between ten and fifteen thousand years ago.

CHAPTER VI

Evolution of Civilization

A more comprehensive investigation into the development of cultural evolution can be found in the book "Wired for Culture": Origin of the Human Social Mind", by Mark Pagel.[3] What follows is a brief summary of those events that marked the evolution of civilization and culture.

Homo-neanderthalensis continued to exist in western Asia and Europe alongside Homo-sapiens after the other members of the Genus Homo had gone extinct. There is now DNA evidence that the two species had interbreed, although to a very limited extent, before Homo-neanderthalensis had gone extinct. The ability of Homo-sapiens to survive on fewer calories, a more diverse diet, and more advanced language skills combined to explain how Homo-sapiens became the only surviving species of the Genus Homo 30,000 years ago. Armed with these abilities, Homo-sapiens, became the representation of humans that not only survived, but migrated to all the continents except Antarctica and learned to live in a wide variety of climates by 15,000 years ago.

All the prior mass-extinctions were followed by the increase of biodiversity as new species adapted to new environments. The evolutionary process has been defined as the process of natural selection. Language and the social consequence has brought about an alternative to natural selection that can be defined as deliberate conscious selection.

Evidence for the origin of this method is apparent in the evolution from hunter-gatherers to herders-farmers, beginning with the domestication of animals, quickly followed by the need to feed those animals when climate conditions would otherwise force them to migrate. Evidence of the first animal to be domesticated is the wolf that in it's domesticated form, became the dog. The dog allowed sheep and goats to be herded. At the same time, Pigs, domesticated hogs, could also be confined provided that they could be feed with scraps, same as the dogs. Without the need to migrate, permanent settlements could be established along rivers that would insure a uninterrupted source of water and structures that did not have to be continually replaced. The next logical step was to develop the ability to feed the ever larger herds of animals during the seasons when their normal source of food was limited. Planting grains and harvesting them for usage during the seasons that they were needed, provided the bonus of supplicating their own sources of nourishment. These developments first occurred in what is now known as the middle-east were the conditions were ideal for the evolution of those animals that could most easily be domesticated and herded. Soon thereafter the water-buffalo was domesticated.

When the permanent settlements became established and grew to the degree that they became targets for those who were not successful in establishing permanent settlements, the need to build a defense resulted in walls and the human resources necessary to guard those walls. Organizing and coordinating those resources required leadership. That leadership represented the first government of what became the first cities. The first archeology evidence of a city is the city of Ur in Mesopotamia with artifacts dating back to 3800 BCE. There are many "permanent" settlements that have not been discovered, some of which are probably older than the settlement of Ur. Even the growth of Ur had many setbacks that destroyed the city as a consequence of floods, aggression from other settlements and earth-quakes. Because of the resources available at that location, the city was rebuilt on top of the debris that resulted in a mound that rose at that location that made it a early target for archeologists. There are many settlements that predate back to the evolution of Homo-sapiens that did not leave evidence of permanent structures. This was because the material used for building the structures were organic and decomposed when exposed to moisture and oxygen. The only remains of these settlements were stones and bones. The markings on these items allowed us to infer how the humans behaved but little on the thought process that lead to that behavior. We know that they learned to weave fibrous plant material into twine and used that twine to attach stone to wood and probably wove plant material into containers for gathering nuts and fruit and other food sources. But not until they collected beads that had no function benefit, and drew pictures on cave walls, did we have any insight on their thought process.

The oldest known cave paintings date to 40,000 years ago. About that time other forms of art began to appear in Africa, Europe, and

Indonesia. Carvings of bone representing Humans, Animals, and composites of the two. Creativity appears to have been a function of population growth rather than genetic evolution because often the creativity disappeared when the population in a specific area shrunk.

The Cro-Magnon Man found in southwestern France, dates between 35,000 and 40,000 years ago, represents the beginning of modern humans. They buried their dead, made bone carvings, developed tools for scrapping leather, fine bone tools, chisel-like tools and retouching techniques that produced tools with flat backs. They lived in shallow caves, Lean-To's against rock walls, or separate shelters made from stone. They appeared to only migrated when forced to as a result of significant environmental changes. They were a robust people and their characteristics can be found though the end of the Stone Age that ended around 2000 BCE.

From an anthological point of view, The stone age is divided into three periods; Paleolithic, Mesolithic an Neolithic. These divisions are determined by the cultural advancements took place, not just chronology of the earth's geology. The Mesolithic period covers the cultural developments that occurred immediately following the end of the last glacial cycle to the introduction of farming. Because the cultural changes took place in different regions at different times, the end of the Mesolithic period overlaps much of the Neolithic period beginning around 12000 BCE to around 3000 BCE and marked by much improved hunting techniques especially with respect to the bow and arrow, the ability to harvest marine animals, and the apparent belief in the spiritual life of human beings. The Neolithic period began in southwest Asia from 10,000 to 6,000 BCE, and from 4,000 to 2,400 in Europe and Africa. Marked by domestication of animals, permanent structures, flint tools, pottery, weaving, beginning of mini empires and writing.

Resistance to migration was enhanced by the ability to trade with distant settlements for resources that were scarce in exchange for resources that were abundant. The demand for permanent structures as trade routes became established, insured that the larger settlements had the resources and bargaining power to demand more for their goods than lesser settlements were able to provide. Walls went up and aggression of one settlement against the other began. When one settlement conquered another, mini-empires were formed. In order for the leadership of one settlement to control another, communication had to be established that insured the new subjects understood a new set of rules or acceptable behavior. This required that language take a new form which would allow the rules to be known without the physical presence of a speaker. Records had to be kept to insure that the rules were adhered to. The solution was the written language. It allowed for the development of empires and the distribution of knowledge. The process began according to currently available archeological evidence, between 4000 and 3000 BCE.

CHAPTER VII

Social Morality

All the religions of the world teach a particular moral code of conduct. The moral code of conduct will be for the benefit of the group though not necessarily for the benefit of the individuals within the group. Sacrificing a member to a God for the benefit of the group is only proceeded by sacrificing animals as perhaps the oldest of all religious practice. The practice continues today through acts of suicide. Suicide for the benefit of the group is an example of submitting to a particular moral code of conduct. The presumed benefit to the individual for making such a sacrifice is dependent on the particular moral code of conduct defined by the specific religion. The act itself is an act of martyrdom which includes suicide bombers.

At the same time that humans were becoming aware of there beliefs regarding mortality, another kingdom was being established on the banks of the Nile river in Egypt that would build the Pyramids and develop another written language, Hieroglyphics. The Egyptian kingdom referred to as the "Old Kingdom" was established when the upper and lower kingdoms became unified under the first Pharaoh in 3150 BCE. The pyramids and the hieroglyphics found in them confirmed that any population center large enough to establish a civilized society, believed in immortality as the ultimate human goal. As the governments of these societies rose and fell, the successes and failures of the government were attributed or blamed to the gods to which they paid tribute. The Old Kingdom came to an end, when the Pharaoh was not able to maintain control of the resources needed to build pyramids, defend the nation, and serve the needs of the majority.

During the evolution of civilization in Mesopotamia, and Egypt, City states were being establishing along the perimeter of the Mediterranean Sea, were able to grow their wealth through the use of ships which was the only practical way of moving people and goods over long distances. On land they were limited to how far one could walk and carry or drag the resources you needed for survival. The Aryan people are credited with the domestication of the horse, applying the wheel to an axle and using the horse to transport material far more efficiently than was previously possible. Much greater distances could be reached in a single day, and when combined with the wheel, allowed armies to transport weapons and supplies making the army much more effective when they reached their destination. Empires could expand over almost limitless territory. The process of domesticating horses took place according to available information, began as early as 4000 BCE in what is now southern Russia and northern Kazakhstan. The chariot was an early application of the wheel an axle and was quickly copied by young Empires in Mesopotamia. Egypt made significant improvements to the chariot in 2000 BCE giving them an advantage on the battle field. The Aryan people migrated from this area into Pakistan and northern India and merged with the Indus Valley Civilization that had a belief system based on the Veda, spiritual poetry handed down orally with no known author, and a message that emphasized reincarnation of one's karma or soul. The Aryan migration became the dominant influence on the Indian language and introduced a written expression of the language known as Sanskrit. The orally handed down Veda poetry was the written down in Sanskrit and new spiritual poetry was added. The development of a class system resulted in a class of people who became those responsible of interpreting the Veda scripture and the intermediary between the people and their Gods. Thus Hindu became the dominant religion of India.

The Egyptian kingdom referred to as the "Old Kingdom" was established when the upper and lower kingdoms became unified under the first Pharaoh in 3150 BCE. The pyramids and the hieroglyphics found in them confirmed that any population center large enough to establish a civilized society, believed in immortality as the ultimate human goal. As the governments of these societies rose and fell, the successes and failures of the government were attributed or blamed to the gods to which they paid tribute. The Old Kingdom came to an end the Pharaoh was not able to maintain control of the resources needed to build pyramids, defend the nation, and serve the needs of the majority.

There were other religions that developed in the sixth century BCE in the middle and far east that challenged the need to go through an intermediary if one wanted to communicate with God. In the far east these were expressed as Taoism and the teachings of Confucius. In

India Buddhism, further west, one based on the teachings of the Greek Philosopher and mathematician Pythagoras. Incite into the cultures that produced these philosophies can be found by reading the book "Creation" by Gore Vidal.[4] The is a fictional story that is based on a mythical person who was assigned by the Emperor of Persia to travel to the far reaches of the Empire to learn the philosophies that motivated their behavior. The book was the only way that Gore Vidal could compare the cultures of such diverse philosophies in one book. All these religious philosophies were established between 590 BCE and 490 BCE which also marks the time that the alphabet had become established as a bases for converting a oral language to a written language. The alphabet promoted the idea that any oral language could be expressed using a fixed set of symbols that allowed people to express their thoughts without devoting a significant portion of their lives learning a vast number of symbols and rules for expressing their thoughts.

There are three religions that define the code of conduct of the vast majority who's seed was the first Book of Moses, Genesis. In Genesis, Abraham's wife Sarah gave birth to his son Isaac, in order to fulfill God's will even though Sarah was not able to give birth and had long past menopause. Furthermore, Abraham already had a son, Ishmael, given birth by slave woman. Ishmael became the patriarch of the Arabs, and Isaac, after Abraham proved his loyalty to God, became the patriarch of the Jews. To prove his loyalty, Abraham had to offer his son as a sacrifice to God. The thought that God required proof of loyalty through the sacrifice of life can only be understood in the context of cultural beliefs at the time that the story was incorporated into Genesis. The writing of the Torah was written with a specific goal: Establish the Jewish people as a unique group of people chosen by God to fulfill God's will on earth. This story established a strong bias that kept the Jewish people genetically separate from the rest of mankind, a bias that continues to this day. The loyalty test insured that Abraham understood Isaac to be not just his son, but the Will of God and patriarch of new ethnically pure group of people.

The Exodus brought new leadership for the descendants of Abraham and those living in the Levant renamed Israel. The coordination of the twelve tribes of Israel activities were originally done by people referred to as Judges. Later the tribes insisted on a king to demonstrate a central authority having the strength of a united nation. The first king was Saul but he was not very effective and was replaced by David, a war hero. David had an affair with the Queen of Sheba and did not set a good example for the Jewish purists. His son, Solomon, showed more wisdom in his administration of the kingdom. He built the first Temple to their God, although the people continued to worship their own personal Gods in their homes as is confirmed by archeological evidence. Solomon also acquired control of mines that brought wealth to Judea. The kings after Solomon were able to use this wealth to build great defensive walls around the larger cities including Megiddo that was referred to as Armageddon in the New Testament.[2] After Solomon the tribes divided into two factions; Judea, lead by the tribe of Joseph with the tribe of Benjamin that was in Jerusalem, and Israel containing the other ten tribes, establishing their Capitol in Shechem but later moved it to Tirzah then built a new capitol, Samaria.[3] Samaria remained the capitol until the destruction of the Kingdom by the Assyrians [4] in 720 BCE ending the kingdom of Israel but not Judea. The conquest of Judea did not take place until 589 BCE when the Babylonian Empire laid siege to Jerusalem that resulted in the destruction of the first Temple. The leadership of Jerusalem, including the Temple Priests were exiled to Babylon. The priests brought with the stories about Moses combined them with stories from Gilgamesh and wrote the five Books of Moses at established the religion of Judaism including the rules and traditions that separated the Jewish people from all other ethnic groups. In 535 BCE, the Persian Empire absorbed the Babylonian Empire and allowed the Jews to return to Jerusalem. On return to Jerusalem, the Jews, with the help of the Persians, rebuilt the Temple and read the Torah to the people which firmly established the religion of Judaism and planted the seed for a manifesto that foretold of a Messiah that would be sent by God to free them from subjugation and establish a "Kingdom of Heaven on Earth" for the Jewish people. The establishment of Judaism has had a profound effect on human behavior not just in the Middle East, but in all of Europe, and the Americas, that continues to this very day.

---

[2] Revelations 16:12,14,16

[3] 1 Kings 16:24

[4] 2 Kings 17:5

Life after death is a belief that dates back to prehistory, but the nature of life after death has taken many variations. As mankind began to develop language skills and began to question the cause of one's observations, through the use of imagination, mankind developed the belief in supernatural entities as an explanation for otherwise unexplainable events. These supernatural entities were given Human personalities but, in order to explain those events that occurred in the past and presumably in the future, the supernatural entities had to be immortal. Immortality became a Human desire and began with belief in Humans having God-like powers, presumably through Humans having relationships with the Gods. The story of Gilgamesh demonstrates the general belief of an after-life as a form of immortality that has become bases of virtually all religions with the possible exception of Atheism. All religions address one fundamental question; What happens to an individual's identity after physical death? Those religions who's core belief was founded on the Book of Genesis believe that, whatever happens, the consequence is for eternity. The alternative is the belief that physical death ends one life cycle and initiates the beginning of another life cycle.

By the time that Christianity had developed as an independent Religion from Judaism, the believers of both Religions became aware that Justice could not be achieved through Human behavior as a member of a "chosen people" or as an individual during a lifetime. If justice was to be archived, that action would need to depend on the action of the individual's behavior during their lifetime. This would be true if you believed in Judgment, presumably by God, or Reincarnation. Even God would need a standard of behavior as a bases for it's judgment and would require knowing how each individual exercised their "free will" in controlling their behavior.

CHAPTER VIII

Age of Reason

All life forms require memory in order to initiate a reaction to the environment. The DNA molecule is the memory of how to make a structure that can initiate a reaction to the sensed environment. In the case of all single cell life forms, all behavior is the consequence of instinctive memory. With the evolution of multi-cellular life forms, their was an explosive diversity in the way life forms could initiate reactions to the environment but the behavior was the result of instinctive memory until the evolution of the animal kingdom. The animal kingdom allowed for coordination of actions between different organisms from a central point that would process sensory information and distribute a response to the appropriate organisms within the multi-cell entity. This central processing center is defined as the brain and required the evolution of the neuron, a cell configuration capable of remembering sensed information. The process of coordination required the coordination of instinctive memory with sensory derived memory. The resulting behavior can simply defined as emotional behavior and describes the animal behavior of all animals capable of conscious thought.

The physiology of conscious thought is the least understood of all definable phenomena's other than the first second of the Big Bang. What we know from observation is that the brain must have a minimal number of neurons that are interconnected in a way that allows adaptive functionality. The actual process of creating thought is the activation of a specific group of neurons in a manner that is not fully understood. What is understood is that the memory of that event does not reside in particular neurons but rather is a particular relationship between groups of neurons involved in creating the thought through input from the senses and memories of similar input. The interaction between the sensed information and memorized information results in new memory and conscious thought. This means that all knowledge is acquired through the senses. In animals, the sensory information travels through the neurons to the brainstem where the instinctive response is evoked and the instinctive memory evokes sensory derived memory from past experiences. The modified emotional response is a secondary response. The primary point is that the emotional response is a modification and not a replacement of instinctive behavior. The degree to which Instinct effects all animal behavior is best explained by E.O. Wilson, a two time winner of the Pulitzer Prize for General Non-Fiction in his book "The Social Conquest of Earth".[5]

The modification of instinctive memory takes place in the Limbic System of the brain and is located directly above the Brain Stem. Organization and functionality of the components that define the Limbic System is extremely complex but there is one key component that is responsible for emotion to sensory memory. The Amygdalate differentiates the sensory inputs measures the relative intensity coming from the various sources and determines which should be given priority. The determination allows memory images that reflect sensory input most in need of attention to be recovered from the individuals sensory memory via the Hippocampus, updating that memory with current sensory input and at the same time issue the signals that prepares the body to initiate the action necessary to alleviate the fear or achieve a desired change. All of this happens as a response to sensory input that initiates and reinforces or modifies one's self-image. Without the Amygdalate their would be no emotional response.

There are clinical examples that clearly show that an inability to make a decision results when one is unable to experience emotion due to a damaged or surgically removed Amygdale. All animal behavior can be explained as an emotional response that describes both the intensity and characteristic of the emotion. The response is even more complex when different emotions are experienced simultaneously but every emotion is the interaction between a component of fear and a component of desire. Frustration and anger are the consequence when the desire component cannot be achieved because of the fear component.

The desire to make a decision is based on emotion but the process of making a free will decision is based on the analysis of past life experiences. Computers are capable of storing a great deal of information and analyzing that knowledge according to a programmed process. What distinguishes mankind from computers is the emotion that created the desire to process the information or the self awareness that can not be programmed. Artificial intelligence is a great tool that can save many problems but their can be no emotional reward for solving any programmed problem. Any action taken by the computer is based upon the desire and self-awareness of the programmers intended action.

The decision making process involves many variables even on simple decisions but free will is not one of those variables, Only Humans are presumably able to use free will in making a decision but the application of free will must only take place after emotional factor has been applied. This would mean that free will is not the cause of emotional decision, but rather an override of the decision. This override is the consequence of reason. The ability to override one's emotional behavior through the use of reason is the essence of free will. The ability to apply reason in the determination of human behavior requires intelligence. Intelligence is awareness, not just of the environment, or of self-awareness, but awareness of the needs of other life forms and one's ability to effect their acquisition of those needs. Understanding what those needs are requires the ability to become aware of other human's thought process's. This awareness was achieved through the development of logic and language. Using this awareness, the degree of fear and desire can be consciously adjusted or overridden resulting in different behavior. Thus, conscious reasoning is the expression of free will. Evoking free will requires a level of awareness that is not present in any animal not capable of understanding the thought process of other individuals. Only Humans have this capability.

Reason does not insure that the behavior is more or less correct, but only that the individual has the ability to determine the criteria that defines what is, from the individuals' perspective, most correct. How one establishes such criteria, is through the process of establishing a self image that defines you as a unique individual. Currently the data suggests that such a self image is not established before the age of five and is constantly being revised as life experiences accumulate. The ability to apply conscious reasoning is not fully developed until a person reaches their mid-twenties. By the time one reaches their mid-twenties, life experiences have shaped your self-image that in turn effects your willingness to apply conscious reason in determining future behavior. People become set in their ways as they get older, unless they become aware of a revelation that causes them to re-examine their self image, stop using their Free-Will.

The development of the Human Being as the product of evolution is increasingly subject to the preponderance of accumulating evidence. In the final analysis, Human life is a continuum of feelings that result in actions. It is not the physical characteristics that define who a person is, but rather the feelings and the ability to understand those feelings that determine how a person will react to life experiences.

From the very beginning of intelligent thought, the cause of phenomenon that could not otherwise be understood was determined to be the actions of supernatural entities, generally referred to as Gods. Millenniums of indoctrination has convinced the vast majority that the act of creation was a supernatural purposeful action. Most religious organizations claim this to be the source but at best, this is second hand information and the truth of this information cannot be verified. If someone says that they had a personal conversation or experience with a supernatural entity, I can choose to believe or disbelieve, but if I base my belief on the experiences of others, then my belief is based on faith, not reason.

Another possible source of this knowledge would be a personal experience acquired by the mind through sensory input. Belief that such a source originated outside the mind's eye is to believe that the experience could not have been obtained by any other means. The question remains as to what initiated this extra-sensory perception. The answer requires the understanding of how reason can be applied to non-physical beliefs.

Indications of intelligence as a desire to understand the cause and effect relationship between the none-physical and the physical universes began approximately 40 to 50 thousand years ago with burials that showed a belief in the afterworld. Because the after-life is a non-physical form of existence, entry to the afterlife requires that an individual must have a non-physical form of existence referred to as one's Soul or Karma in many of the so-called eastern religions. A non-physical form of existence is certainly possible, thought itself is a non-physical form of existence that originated from physical interaction. The Soul could as easily dissipate into the non-physical universe just as the physical components of the body dissipate into the physical universe but that would require that the Soul or Karma had no purpose.

Religion attempts to define the laws of the non-physical universe, science attempts to define the laws that govern the physical universe.

Both use the power of reason to make their case but science has the distinct advantage of the quantitative language called mathematics. Truth is what happened and science requires quantization of any phenomenon through mathematic measurement in order to be accepted as eligible. Since non-physical events cannot be quantized, only the consequence of such events can provide evidence that supernatural events initiated physical events. Such events could be the consequence of evolution in the supernatural universe or the intended action of a supernatural entity. Either way, the Soul could continue to evolve after physical death. Extra-sensory perception is subjective evidence that supernatural existence can represent entities once attached to physical entities. Awareness of this connection is most often the result of the imagination that was created in the mind from memory that was formed internally or externally. Verification that the memory was the externally induced is currently impossible. The only evidence that could provide a high probability of an externally induced memory, is that the individual acquired knowledge that had no other plausible explanation. there are a sufficient number of such awareness that make extra-ordinary perception the consequence of externally induced memory. If supernatural entities are able to induce memory, than they must have spiritual memory and this would apply to the Soul. The purpose may well have nothing to do with unverifiable scripture that was written with a bias.

Burial of the dead required a belief that a spiritual representation of physical immortality could be achieved at least by those who had a relationship with the Gods. Physical assets of the individual could transcend along with the body to the underworld, representing a kind of physical immorality. Reason was applied to justify the belief but the bases of the reasoning was faith in Gods being the only entities that could grant immortality.

CHAPTER IX

Judgment

Reason allows Human Beings to understand that human behavior can be controlled, changing one's beliefs in cause and effect. The self-image that was the result of sensory input could be controlled by the establishment of social standards of right and wrong. These beliefs were subject to manipulation by individuals that had the greatest influence on the society. They did so by establishing a belief based on the mythology that only Gods could control the forces of nature and only individuals chosen by the Gods could influence the Gods behaviors. Those able to convince the society that they were chosen by a God were able to change the behavior of those citizens that were convinced. Those not convinced were often punished for their non-belief by acts of nature, presumably initiated by the Gods, or actions taken by the governors of the society. The result was that the Individual could not insure that justice could be assured during one's physical lifetime regardless of the individuals behavior, but was determined by events beyond the control of the individual.

Burials of kings and emperors began to include more assets that would insure that they would an immortal existence of a God. The pyramids of Egypt carried this practice to an extreme. Early in the establishment of the New Kingdom, an edict proclaimed that the father of the Pharaoh was the physical incarnation of a God and therefore his children and their descendants where the only persons eligible to become Pharaoh.

The belief that one's position in the life after death is dependant on the closeness of one's relationship with God is wide-spread and continues to this day. In the Judeo-Christian and Islamic religions, not having a relationship with God prevents one from entering the afterlife. If God must choose to have a relationship with you, then what is the bases for God to make that choice? It can only be based on observation of one's behavior. If an individual or group of individuals was not aware of the existence of God, then by what standard should their behavior be judged? These are the dilemmas that make people question the premise that "By the grace of God one is saved". Why would God create Human Beings, give them "Free-Will" and then punish them for exercising their Free-Will in a manner that God considered inappropriate. How can such a God be considered Just. These are the reason that many people refuse to accept either God as the creator or God as the Judge.

There is another belief that allows the individual to achieve immortality that determines one's position in the afterlife based on one's behavior; Reincarnation. The belief in reincarnation and practiced as a religious belief, dates back to the Indo-Aryans of northern India around 1800 BCE. Upanishads are Veda texts concerning the nature of reality (Brahman), and describing the path to human salvation. Brahman is gender neutral and referred to as the supreme self. Para Brahman (beyond self) or Nirguna-Brahman (without form or qualities), represents the absolute truth. Transcendentalists call this Brahma, Paramatma, or <u>Bhagavata Purana</u>. Whoever realizes the truth of Supreme Brahma, attains to supreme felicity (happiness, joy, bliss). In order realize this salvation, one must separate from the supreme self and acquire a false personality called a jiva (living being). Due to egotism, desire motivated actions, and selfishness they experience suffering, caught in a physical world that has no escape except through reincarnation of the soul and then only to another life form. Based upon lessons learned and deeds performed, the soul either progresses or regresses in it's quest to escape from the mortal world. Hinduism allows the soul to go through many life cycles that may include a multitude of both regressions and progression until the soul escapes the immediate return to another life cycle in a state of hell or heaven. In either case their stay is not permanent, having received their reward or punishment they return to the physical world as rain where they are absorbed by plants, when the plants are eaten by an animal, they become semen that enters the womb of a female until the female gives birth. A new series of life cycles begins and the process continues indefinitely.

Buddhism is not very well understood in the West, but there are striking similarity with Christianity as taught by Jesus of Nazareth. Both Buddha and Jesus taught and practiced selflessness. Buddhism teaches that desire inevitably leads to sorrow and to achieve lasting happiness is to behave in a manner that is selfless. Such behavior is extremely difficult to achieve and impossible to achieve in a single lifetime. Buddhism neither required nor denied the existence of Gods but taught that your destiny is not dependent upon outside influences either physical or spiritual. As long as you are able to control your desires, you can determine your own reality and consequently your destiny. By denying outside desires, you can achieve freedom while imprisoned physically or psychologically.

According to believers in Jesus, his death marked the beginning of life after death. The nature of life after death was totally dependent on one's religious philosophy and how that philosophy was enacted. If that religious philosophy is to act in a selfless manner for the benefit of others, then Buddhism and Christianity are compatible philosophies. The reward for living a totally selfless life is Nirvana in the first case and Heaven in the second case. The consequence of living a life that is not selfless and does not recognize the authority of a predefined God, is where Christianity differs from Buddhism. In the case of Buddhism, you are reborn into a new lifecycle, In the case of Christianity, you are confined to an eternal state of spiritual existence In one case the soul, is a birthright, in the other the soul of every individual is bestowed by a supernatural entity. Jesus never taught that heaven or hell were eternal events but that was the implication.

There is evidence that Pythagoras believed in reincarnation but the evidence is based on second hand reports that were accounts of those that were followers of Pythagorean philosophies. Presumably Pythagoras could remember his past lives. The Pythagorean philosophy originated around 500 BCE but knowledge of the practice was not written until the time of Socrates, Plato, and Aristotle.

Knowledge of religious practices in the Far East were not revealed in writing until several centuries after the fact, not because they were not written, but because those writings were destroyed. From what can be deduced from the writings about those practices, the people that would eventually become defined as Chinese, practiced ancestor worship. The belief evolved into the belief that every individual had two souls, a spiritual soul that transcended to immortality when one died, and a soul of the body that remained bound to the physical world needed to be cared for in order to insure that the body soul would protect the family or at least not do harm. The Ruler was given Heavens Mandate to rule so long as the Ruler took care of his people. Defeat by another Ruler meant that the mandate was lost. Confucianism introduced the belief that the fundamental nature of Human Beings was virtuous and that the responsibility of the ruling class was to insure that the citizens were treated fairly. The belief in reincarnation did not occur east of India until Buddhism was introduce in 65 BCE. Buddhism was soon exported to the Korean peninsula, South-East Asia, and later (by 552 CE) to the Japanese Islands. The most popular form of Buddhism was Zen Buddhism which incorporated Taoism (Daoism). Even so, their were periods of resistance and persecution because Buddha was a foreigner.

By 800 CE, Cultural bias had divided the vast majority of Humanity into two camps regarding life after death, Those that believed their creator was God who determines what happens to their soul at the end of their physical life, and those who believe that their behavior determines the form of reincarnation that will take place in the next physical life cycle.

Life experiences define you and are the bases on which your life is judged by others when you die. If you are judged, then by definition you are being held accountable for your actions. This would imply that there are actions that can be scaled between right and wrong and that is the bases of morality. The paradox is that if we are held responsible, then we must know what actions would be judged as wrong. Unless we are born with this knowledge, we cannot be held responsible for our behavior.

God could judge souls as one might judge a harvest from a garden, keep the best and toss the rest. Biblical scripture supports this point of view by suggesting that worshiping God not only insures your life after death, but improves your position in the hereafter. Although such behavior is selfless, it is self destructive and totally dependent on a reward bestowed by God after death. It is unreasonable God intended that free will be used in a self destructive manner. Selfless behavior is the highest moral standard but true selflessness should never be self destructive. On the contrary, accumulation of wealth for the benefit of others trumps accumulation of wealth for the satisfaction of personal desire, just as love should be for the benefit of the one you love and not for the satisfaction of your desire to poses.

What we seek are universal truths to guide us in making these decisions which often conflict with our cultural bias. The history of evolution is the only universal truth that we have access to, limited as it may be. By studying the evolution on Human civilization we can become aware of those decisions that were most beneficial to the Individuals and their societies. Through the study of history we can see that there are principles that should allow us and perhaps require us to make the sacrifices including death, for the benefit derived from advancing a belief in a principle. How to determine if the principle is worthy of our sacrifice is the dilemma facing each and every Human Being capable of applying reason to the decision making process. There is no universally correct answer to any dilemma if for no other reason than that every individual's situation is different and their level of awareness limits their ability to apply reason. Furthermore, whatever would be the best decision at one point of time would not necessarily be the best decision at another point of time when the situation would have inevitably changed. Developing the principles that will guide one's philosophy is a life long endeavor that requires the continuous acquisition of knowledge. Seeking out a social group that allows one to pursue one's principles as they develop and supports one's acquisition of knowledge, should be your guide in developing your social environment. Every individual having the ability to reason may utilize their free will, but as a result of social conditioning, may suppress their free will. Participating passively allows you to function in a particular culture with minimal conflict. As long as such participation does not conflict with your principles, you will not feel guilt. The greater the degree that the moral standard of a particular social group matches your own personal standard, the more actively you will participate in the activities of the group.

I disagree with established Christian religions self proclaimed authority to establish the rules by which Christian morality shall be practiced. I could say the same for all non-Christian religions, even though I do not have the cultural experience that I have had with Christianity. Each individual is a thread in the fabric of life. Each persons effect on people is what gives that person's life meaning, good or bad and therefore the primary goal of society should be to insure that all it's members be given the opportunity to maximize their abilities for the benefit of both the individual and the society. Immortality, even if achievable should not be the primary goal and certainly not be determined by the grace or judgment of God or any other external entity. Such judgment would render the free will achieved through reason meaningless.

CHAPTER X

Evolution of Cultural Bias

Part 1 (Religious Bias)

Only through writing can we have the understanding of the psychology that defined the cultural behavior during the Ancient, Classical, and Middle Ages. Ancient civilization, is a time period marked by the earliest written languages, hieroglyphics and cuneiform, from the third millennium BCE to the sixth century BCE, when all the major philosophies from around the major population centers had been defined. The Classical Ages mark the time that the literate became aware of those philosophies through distribution of hand copied literature, (500 BCE to 500 CE). The Middle Ages is marked by the fall of the Roman Empire to the awareness of the physical size of the planet Earth. (500 CE to 1500 CE). The groundwork for establishing Cultural Bias based on Religious beliefs is found in the oldest story ever recorded when ancient cities were uncovered in Mesopotamia and nearby regions of the middle east by archeologists. These cities had large buildings used for keeping records of transactions regarding doing business anywhere in the empire. The records were kept in a large building that not only contained business transactions, but declarations, treaties, etc. and a story about a king that had ruled more than a millennium earlier. Everything was written on clay tablets and survived the destruction of the buildings. There were many copies of the story and several versions. All had to be pieced together and decoded from an ancient language written in cuneiform (wedge marks). The story was compiled into one book that consolidated the different versions into one book that has been published under the title "The Epoch of Gilgamesh".[6] The tablet fragments dated back to 2100 BCE, about King Gilgamesh who ruled over the city-state of Uruk on the Euphrates river around 3000 BCE. The story is a mythology that like all mythologies is based on a few seeds of truth that are covered by many half-truths, exaggerations and fictional accounts. The central theme of the story is Gilgamesh's desire and attempt to achieve immortality. What the story reveals is that by 3000 BCE, humans believed in immortal

Gods who were responsible for events that had no reasonable explanation and that they were aware of their beliefs.

These Gods had human personalities and used their powers to create and manipulate humans for their own pleasure. These beliefs were not the result of specific event, but evolved along with their physiological evolution of hunting, gathering, and language skills. In summary, they became aware that life is not fair and that justice could only be achieved through immortality.

There is another revelation that is apparent in the story of Gilgamesh that requires one to be familiar with another mythology, the bible or more precisely "Genesis" the first book of the Torah attributed to the story of Moses. The Torah was written by Jewish priests who were exiled from the temple in Jerusalem to Babylonia when Judea was conquered by the Babylonian Empire in 589 BCE. The Torah was completed and read to the Jewish people when they were allowed to return to Jerusalem in 539 BCE. While the priests were exiled, they certainly would have become familiar with the story of Gilgamesh. The story of Gilgamesh includes the story of "Noah" including the building of the arch, loading it with pairs of all the animals, coming to rest on a mountain top, and sending three pair of birds to find a place to nest. Since the story of Gilgamesh had been written many centuries before the story of Noah, there can be little doubt of the source from which the story was derived. There is another place in the story of Gilgamesh where Gilgamesh is informed of a plant that if eaten, would provide him with eternal youth. The plant, once obtained was eaten by a serpent and Gilgamesh had to leave the land of paradise without having achieved immortality. This most probably inspired the story of Adam and Eve. During the time of Gilgamesh, a religion was established in the kingdom of Babylon based on the teachings of a self proclaimed prophet Zoroaster known as Zoroastrianism that introduced the concept of good and evil as the will of two Gods being imposed on mankind.

Cultural Bias was a development that occurred symbiotically with the development of civilization. Our knowledge of cultural bias is based on archeological discoveries and text written in ancient languages that allows us to understand what these people were thinking. The various cultures developed their beliefs based on stories handed down from one generation to another. Until these stories could be written and preserved, there is no way of knowing what stories were lost or what changes took place as the stories were repeated through the generations. Even those that were written were often destroyed by cultures that replaced them. The written documents represent symbolic representation of oral speech that is often difficult to interpret. The writings reflect events that had significant effects on cultures that wrote those stories. The institutions most directly effected were the governing institutions. The events often defined natural events, usually catastrophic, and individuals who's actions either caused, predicted, or responded to such events. The first need of any civilized society is to defend and preserve the society. Any time there are two or more people, there will be conflict of interest. This conflict may be resolved by agreement that requires the understanding of each others position for compromise, lack of understanding on one side or the other about the others position leading to an unfair compromise, or violence as a last resort. From the family to nations, leadership will favor the male and therefore males will always dominate the leadership roles from Emperors down to the patriarchal family. Therefore gender bias is the oldest of all cultural biases and permeates all social activates. The second greatest cultural bias based upon ancient writings from the earliest civilizations are with regard to religious beliefs meaning beliefs that have no verifiable bases on which to establish reasonable belief. The bases of such beliefs is faith in those who initiated the story that supports the beliefs. Because such beliefs have had a major effect on Human behavior, this Chapter will concentrate on the development or religious cultural bias.

The vast majority of Western Religions can trace the origination of the ideology back to Egypt and the Levant or Eastern Region at the east end of the Mediterranean sea. This ideology has expanded to account for the beliefs of nearly half of the earths Human population. The cultural bias that became so influential begins with the hieroglyphic script chiseled into stone describing the beliefs and accomplishments of Pharaohs of the New Kingdom. All cultural changes are initiated by individuals. The degree of influence that they have on any society depends on how many people are effected. If the society grows into an Empire, then the cultural beliefs will be exported and melded with local cultural beliefs. Those individuals who initiated ancestor worship through burial rituals, are lost to history as well as those who initiated the worship of immortal Gods.

By the third millennium BCE, the southern Levant was a land of small, fortified towns and villages. A major trade route connecting Mesopotamia with Egypt ran south from Damascus through the Jordan valley. Late in the third millennium, Amorites, a nomadic people who depended on herding, were moving into northern Syria, and their close relatives, the Canaanites expanded westward to the coast of the Mediterranean Sea, some entering and taking control of northern Egypt. The Canaanites established an independent settlement called Goshen, by 1720 BCE. Subjective evidence strongly suggests that Goshen was a Canaanite settlement later to be designated as the first Israelite settlement and was located near the fortified frontier city of Zarw. Much of the fortifications were made of mud bricks known only to Israelite masons. Asiatic invaders known as the Hyksos, Took control of the entire delta region of Egypt in about 1650 BCE and ruled northern Egypt until they were expelled an all of Egypt was once again reunited beginning the New Kingdom in 1549 BCE. The Hyksos brought several technological advances to Egypt, including the horse and chariot, the composite bow, improved battle-axes, improved fortification techniques, techniques for working bronze and pottery, plus new breeds of animals and crops. These advances allowed the New kingdom to expand Egypt into the world's largest Empire.

CHAPTER XI

Evolution of Cultural Bias

Part 2 (Judaism)

Dating events and locations described in the available copies of the original Hebrew Bible is difficult, not only because of the changes introduced in the acts of copying and translating, but because these versions make no reference to any Pharaoh by name or identifies any event that can be collaborated with Egyptian accounts of such events or their locations. Although subjective, evidence strongly suggests that Goshen was the first Israelite settlement and located near the fortified city of Zarw. Zarw. Zarw became the location that the Pharaoh Amenhotep III and Queen Tiye established a royal residence, Zarw-kha, within the walls of Zarw. Queen Tiye was the daughter of the Vizier Yuya who married one of the Pharaoh's daughters making Queen Tiye of Royal blood. Evidence suggests that Yuya was first named Vizier by the Pharaoh Thutmose IV because of his ability to read dreams, linking him to Joseph, Abraham's eleventh son.

There are events recorded in written history that do identify the individual responsible for the initiation of cultural change. The first person to have a significant effect on a societies cultural beliefs and consequence behavior was the Egyptian Pharaoh Hatshepsut who was the eldest daughter to Thutmose I. After her father's death in 1493 BCE, Hatshepsut married her half-brother Thutmose II and became the queen of Egypt. Thutmose II died in 1497 BCE and the throne went to his infant son Thutmose III. Hatshepsut acted as regent presumably until Thutmose III became of age. However, in less than seven years Hatshepsut assumed the title and full powers of the Pharaoh becoming co-ruler of Egypt with Thutmose III. Hatshepsut defended her legitimacy claiming that her father had appointed her successor and with the help of the temple priests, established the belief that her father was the God Amun incarnated as a Human. That made Hatshepsut the divine daughter of God. This allowed her to take on the appearance of Pharaoh including the false beard when acting in the official capacity of Pharaoh. When Hatshepsut died, Thutmose III had all depictions of

Hatshepsut removed but the belief that one had to have kinship with the descendants of Thutmose I to become Pharaoh became part of the cultural tradition that became a significant factor in who the next several Pharaohs would be.

The sequence of events that followed are described primarily in the bible and are debated among Egyptian and Christian historians. What is not debatable is that in 1350 BCE Amunhotep IV became Pharaoh. By this time there was a very strong relationship between the Pharaoh and the priests of the various temples dedicated to the various Gods. On becoming Pharaoh, Amunhotep IV declared that there would only be one God, Aten the Sun God. He changed his name to Akhenaten and had a new Capitol city constructed with one Temple dedicated to the God Aten as the centerpiece. He then had the entire population of Thebes moved to the new city called Armana. There has been no reasonable explanation as to why the Pharaoh made such a drastic and fundamental change to monotheism that earned him many enemies until Ahmed Osman, an Egyptian journalist who was raised a Muslim, converted to be a Christian, and learned Hebrew, Arabic, and Greek, Wrote two books in an attempt to reconcile the biblical stories with Egyptian history. The first was "Stranger in the Valley of the Kings"[7] that explained why a tomb was discovered that contained the mummy of one that did not have royal blood. The short conclusion was that it was the body of Joseph, Abraham's son who was sold into slavery by his brothers but became the trusted advisor to a Pharaoh and was allowed to marry the Pharaohs daughter. The second book was "Akhenaten and Moses"[8] that provides a detailed explanation as to why the two are one in the same. A succinct synopsis of the book can be found at the website of Graham Hancock[5]. The book clearly establishes the most probable timeline of the Akhenaten-Moses exodus taking place during the reign of Ramses I (1335-1333 BCE). This scenario would also explain how and why the people of Israel were aware of the religious practices that took place in Armana.

Because of the destruction and suppression of written information, knowing what brought about the cultural changes that took place over the next several hundred years has been left to interpretation of the scant evidence available and speculation. There is information that

---

[5] http://grahamhancock.com/moses-akhenaten-same-person-osman/

describes events that took place in Egypt, but was written long after the events took place and were in themselves biased versions of the events. The most often referenced document is the first five books of the bible or more specifically the Torah written by the Priest of the Temple in Jerusalem while they were in exile in Babylonia. Even this text had been hand copied many times with the original versions being lost before versions exist that have been preserved. When the Persians allowed the Temple Priests to return to Judea, they brought with them the Torah,

A copy was taken to the Samaritan community of Judea. Over the centuries the text was copied, The Samaritan copy was written in the Samaritan alphabet sometime between 432 BCE and 122 BCE. According to an ancient historian Josephus, more specifically during the time of Alexander the Great. The Samaritan text became known as the Samaritan Torah or Samaritan Pentateuch which differs from the Jerusalem or Masoretic Text copies mostly by minor variations in the spelling of words or grammatical constructions, but there are two significant semantic differences. One is that the Ten Commandments contain a Command that a Temple be built on Mount Gerizim, not Jerusalem. The other significant difference is in Exodus 12:40.The Samaritan Pentateuch states: "Now the sojourning of the children of Israel and their fathers which they had dwelt in the land of Canaan and Egypt was four hundred thirty years." This agrees with the Septuagint translation of the Hebrew Text translated into Koine Greek in the third century BCE by seventy Jewish rabbi scholars. The Masoretic Text does not include the word Canaan. The significance is aligning the events in the Torah with events known to have taken place in Egypt.

In his book "Stranger in the Valley of the Kings", Ahmed Olson points out that the events in Egypt and those implied in the Torah can be aligned much more reasonably if misinterpretations of the timeline are corrected. It also explains cultural changes that otherwise have no reasonable explanation. According to the Hebrew Bible during the time of famine in the land of Canaan, Joseph, 11th son of Jacob, was sold into slavery to the Egyptians. The land of Canaan was so named after the son of Abraham who established a tribe that spoke Hebrew, a variation of the Aramaic language as did the surrounding tribes that had a mixture of ethnic backgrounds. Hebrew speaking people fled the Levant and established a community in Goshen on the eastern edge of the Nile Delta, a land suitable for both crops and livestock. What happened to

Joseph had a profound effect on the Egyptian culture and future cultures throughout the Mille East that spread across North Africa, Europe, and the Americas. Until recent history their was no physical evidence to support the stories of the Hebrew bible. As Archeology has uncovered physical evidence that supported the possibility that some events in the Hebrew Bible correlate with events that took place in Egypt.

According to the Hebrew bible, a son of Abraham referred to as Joseph, was sold into slavery to the Egyptians but rose to become Vizier, the second in power only to the pharaoh. Physical evidence though circumstantial, strongly suggests that a person from the land of Canaan was sold into slavery and could have become a Vizier during the Eighteenth Dynasty possibly during the reign of Tuthmosis IV (1413 -1405 BCE).[6] In 1905 the tomb of a Vizier of non-royal blood was found in the Valley of the kings" but was dismissed as an anomaly until in depth analysis of the Vizier's Tombs continents and other information that narrowed the time that he held the position. One of his titles matched the title given to Joseph in the Hebrew Bible. Along with other similarities between the two strongly suggest that the two were the same person. Yuya was the name on his tomb and was allowed to marry one of the Pharaoh's daughters making his children of royal blood. The Pharaoh succeeding Tuthmosis IV was Amenhotep III but was too young to rule for the first twelve years of his reign and Yuya continued as Vizier in all probability until Amenhotep III was able to rule without a co-regent. Amenhotep III took as a wife one of Yuya's daughters making any son from that wife eligible to become Pharaoh. Yuya rose to his position as a result of his ability to interpret the dreams of the Pharaoh. He certainly would have suggested to the Pharaoh that there was one supreme God that was not the God Amun as suggested by the Pharaoh Hatshepsut. This would have been extremely threatening to the Temple Priests and would have put Yuya and his families lives in danger. It is understandable that Yuya had his wife and children removed from Thebes to a royal residence to the far north. This would have included the daughter that became Amenhotep III's wife who's son would be the next Pharaoh. The royal palace would have been near the settlement established by Yuya's father's descendents including

---

[6] The circumstantial evidence is clearly presented in the book "Stranger in the valley of the Kings"e

Yuya's brothers. Yuya's father was Jacob who was given the name Israel according to the Hebrew Bible and his descendents who settled on the border to Egypt were referred to as Israelites. Yuya would have been informed that his brothers had returned there to trade with the Egyptians along with the youngest brother, Benjamin. The story of the Vizier negotiating with his brothers could easily have taken place. With Yuya's daughter and Amenhotep III's wife giving birth to a son, the temple priests would have felt threatened and sent agents to kill the child. Although the Palace and settlement has been destroyed and the ruins have yet to be identified, circumstantial physical evidence of these events along with the birth and life of a person referred to as Moses in the Hebrew Bible.[7]

A story of the Pharaoh's wife putting her son in a raft made of reeds and sent down the river to the protection of her Israelite relatives, from the order by Tuthmosis IV to have the midwives kill Queen Tiye's child if it turned out to be a boy. This was to insure that the God Amun would not be replaced by the Primary God of Zarw, Aten (Sun God) if Tiye's child became Pharaoh. The midwives did not fulfill their required task and Tuthmosis IV ordered that the first born male of all Israelites be killed. The order was rescinded when he was told that Tiye's child was thrown into the Nile. Queen Tiye arranged to have her son marry his half sister Nefertiti- daughter of Amenhotep III and legitimate heiress, appointed his father's coregent, emphasizing Nefertiti's role in order to placate the temple priests. When Amenhotep IV came to Thebes as coregent with his father, he built temples to his monolithic God Aten at Karnak and Luxor. When Amenhotep III died in 1367 BCE, Amenhotep IV realizing that the opposition in establishing Aten as the Primary God in Thebes was not achievable, founded a new capitol at Tell el-Amarna, 200 miles to the north. The cultural changes that were fostered in Armana can be illustrated by the tablets that were uncovered when archeologists uncovered the ruins of Armana. The tablets contained records of transactions between Egypt and those living east of Zarw. The tablets were written in cuneiform, the language that originated in Mesopotamia and reflected strong cultural ties with the Israelites, the nearest eastern settlement of Goshen.

---

[7] The evidence that substantiates the correlation Yuya's life with these events described in the Hebrew Bible can be found in the book : "Akhenaten and Moses"

According to the translations of the Hebrew bible, the population of the settlement exploded from the original seventy including Moses and his children to over 600,000 males plus their dependants. Such an explosion in population could only have happened if Egyptians who became believers in the monolithic God of Aten migrated from Egypt enhancing the population of Goshen. This would also explain why those Egyptians who maintained their allegiance to the God Amun felt so threatened by the actions of Amenhotep IV who had changed his name to Akhenaten that they forced Akhenaten to appoint his brother, Sememkhkare, as coregent at Thebes. When Akhenaten is forced to abdicate, Sememkhkare becomes Pharaoh but only for a few days when he replaced by Akhenaton's son Tutankhamun-"living image of Amun", originally Tutankhaten-"living image of Aten" but changed soon after becoming Pharaoh. Tutankhamun. Was nine or ten when he succeeded Akhenaten. Aye, too young to rule Egypt on his own, his close advisors and effectively coregents were Aye, presumed son of Yuya and Grand Vizier of Akhenaten, and General of the Armies, Horemheb. Without the support of the Army, the majority of which believed in the God Amun, replacing the God Amun with the monotheist God Aten was an impossible task. The consequence was that Akhenaten was forced to abdicate and allow Sememkhkare to become Pharaoh of all Egypt who would allow the Gods Amun and Aten to co-exist. Sememkhkare died, probably assassinated, within a few days and was replaced by Tutankhamun who strongly influenced by his coregent Horemheb, rejected the tile Tutankaten (son of Aten), and returned to being the son of Amun. The worship of Aten was not banned, Aten was just one of many God's subordinate to The God Amun. Tutankhamun died at the age of eighteen and has referred to as the Boy King. Aye, presumed son of Yuya, who was an old man by this time, married Tutankhamun's wife against her will, an thereby secured the position of Pharaoh allowing the worship of Aten to continue.

Tutankhamun's wife had refused marriage with Horemheb which made him feel that he had been cheated out of becoming Pharaoh. Four years into Aye's reign, Horemheb married the sister of Akhenaton's wife Nefertiti and seized power declaring himself Pharaoh. He immediately took over all the monuments to Aye and Tutankhamun and totally destroyed the temples to Aten, which he hated. Horemheb also turned Zarw into a virtual prison for those who continued to believe in

Aten and most probably provided the labor to build a replacement frontier Palace and fortress to be known as Pi-Ramses. He took over the mortuary temple of Aye and began dating his reign from the death of Amenhotep III eliminating all references to Akhenaten. Horemheb, having no children, appointed a person Paramessu, who was of non-royal birth, as vizier and High Priest of Amun. On the death of Horemheb, Paramessu became Ramesses I - "Ra bore him" establishing the nineteenth dynasty and designating his son Seti as the crown prince. Ramesses I had a short reign and is an excellent candidate for the Pharaoh in power when the bible suggests that Moses made his exodus from Egypt after his failed attempt to reestablish his position as Pharaoh.[8] Pi-Ramses was the city from which Akhenaten/Moses lead the exodus fro Egypt to the Sinai. At a temple dedicated to Seti found at Abydos, Egypt, a list of all past Pharaohs was carved into a limestone wall. The list excluded those Pharaohs considered to be illegitimate which included the four Pharaohs referred to as the Armana Pharaohs - Akhenaten, Sememkhkare, Tutankhamun, and Aye. The difficulty of aligning events in the bible with events in Egypt during the time of the Armana Pharaohs became extremely difficult. New discoveries and reinterpretation of old discoveries have begun to overcome past prejudice and provided those who are able to recognize past prejudices with an insight on how prejudices are created, resulting in behavior that is often counterproductive to civilization. Depending on who's timeline you accept, the Exodus took place between 1333 BCE and 1296 BCE during the time of Ramesses I and probably was the cause of the Pharaoh's demise.

The Hebrew Bible seems to have numbered the people who followed Moses into the Sinai as 600,000 males plus their dependents which would have totaled conservatively at over a million. This was at a minimum a gross exaggeration for many reasons and most probably numbered a few thousand. Even they would not have all been Israel descendents.

The timeline for the time spent in the Sinai is more difficult to establish because there are no events that took place in the Sinai that can be aligned with events that took place in Egypt except the birth on Moses/Akhenaten which was about 1394 BCE. If Moses left Egypt

---

[8] See the book Moses and Akhenaten by Ahmed Osman

during the second year of Ramesses I, about 1333 BCE or at the age of about sixty. Even if those who followed Moses into the Sinai numbered a thousand or so, they would have need an area in the Sinai that would support that number. The corridor from the frontier fortress in Egypt to the land of Cannon, the gateway to Asia, would be patrolled by Seti's army leaving only South Sinai available to Moses. Most of the southern area of the Sinai is scrubland but there was a settlement established for mining Turquoise called Sarabit, not far from Mount Sinai were Moses went to speak with God and received the Ten Commandments. On the high peak of Sarabit there was a Temple Shrine. Fragments of a limestone stela was found where Ramesses I described himself as "the ruler of all that the Aten embraces". There was evidence that the rituals preformed at the temple were of Semitic nature and a statuette of Queen Tiye, Akhenaten's mother. This is where Akhenaten came to the realization that Aten and Jehovah[9] were one and the same.

It was after returning from Mount Sinai with the Ten Commandments that Moses decided to leave the Temple and complete his journey to the "promised land". This required going north across the Moab desert and crossing the corridor between Pi-Ramses and the land of Cannon. This is when Moses uses his scepter to provide water for his followers. According to the copies of the Hebrew Bible, God punished Moses for using his scepter to provide water to his people but why would God punish Moses for doing something that was not explicitly forbidden? A far more likely scenario is that Akhenaten used his Scepter as a sign of his authority in obtaining water from one of the fortresses between Zarw and Gaza built around wells. Seti I would have been informed which would have resulted in a confrontation between Seti and Akhenaten on the top of mountain in Moab where Moses presumably died.

In 1336 BCE, Akhenaten supposedly died under circumstances that are questionable and Armana was abandoned. Based on more recent data uncovered by archeologists, Akhenaten did not die in 1336 BCE but went into a self imposed exile in northeastern Sinai were he gathered his supporters and attempted to retake his place as Pharaoh. When he failed, the exodus that was told in the bible took place but probably amounted to no more that a few thousand people. Many others, who

---

[9] see Exodus 6:3

were followers of the new monotheist religion and Loyal to Akhenaten, exited south along the Nile ending in what is now northern Ethiopia and southern Eritrea, an area that later became the domain of the Queen of Sheba. The biblical account of the exodus was at best grossly exaggerated, and at worst a complete fabrication.

There is evidence of a migration of people from Egypt to the land of Cannon and that they provided the leadership that established the twelve tribes of Israel as a loose federation but most of the people who formed those tribes, never left the land of Cannon and were descendents of Abraham. Joshua, Moses' second in command would then have crossed the Jordan river into the promised land sometime around 1300 BCE or no later than 1270 BCE. Once established, Joshua designated the territories of the twelve tribes as being decedents of Abraham. Based on Joshua's probable age at the time he entered the promised land, the establishment of a loose federation of the twelve tribes of Israel with Judges coordinating common defense efforts among the federation tribes must have completed by 1200 BCE. The so called Judges indoctrinated the tribes with belief in their common heritage. When the Philistines defeated the Israelites and captured the Ark of the Covenant, the Israelites became convinced that they needed a King to unit the twelve tribes into a single Kingdom. The first King was chosen by Samuel, the last of the Judges and the first of the major Prophets. Saul who was from the tribe of Benjamin, took the position of King presumably under protest but once King, he took advantage of the position and was forced to step down briefly being replaced by his son but was soon replaced by David who was a war hero having slain Goliath, a fierce leader of an opposing Army.

The authors of the original Hebrew Bible exalted King David not only because he defeated the Philistines and recovered the Arc of the Covenant, but was able to capture control of Jerusalem, thus establishing Jerusalem as the Capitol of a united Israel called Judea. Jerusalem also gave the Israelites access to, if not outright control of the copper mines just east of the city, the resources to begin a building campaign that began with David's son, Solomon building a Temple for the Arc of the Covenant. Much of the rebuilding of cities was credited to King Solomon but recent advances in dating ancient ruins has revealed that the rebuilding took place during the hundred years following King Solomon's reign.

There was a dispute as to were the Temple should have been built. A copy of the original Hebrew Bible was taken and a new settlement and became known as the Samaritans. The Samaritans were people who lived in what had been the Northern Kingdom of Israel. Samaria, the name of that kingdom's capital, was located between Galilee in the north and Judea in the south. The Samaritans were a racially mixed society with Jewish and pagan ancestry. Although they worshiped Yahweh as did the Jews, their religion was not mainstream Judaism. They accepted only the first five books of the Bible as canonical, and their temple was on Mount Gerazim instead of on Mount Zion in Jerusalem. They developed their own alphabet and translated the Hebrew text probably during the time of Alexander the Great and has then been copied faithfully over the centuries. Many of the differences are semantic and of no significance, but there are some that significantly change the interpretation of the text. One of those defines where the Temple should be built. Another redefines the time that the Israelites spent in Egypt. These differences as well as other differences between the tribes who's origin was from Egypt, the Joseph and Benjamin tribes, and those tribes who never left the land of Canaan, the other ten, caused them to separate into separate kingdoms. Judea retained Jerusalem as its Capitol and the northern kingdom that took the name of Israel. Shechem was the first Capitol of Israel, later it was moved to Tirzah. King Omri built a new city in Samaria[10], which continued as such until the destruction by the Assyrians[11]. During the three year siege of Samaria by the Assyrians, Shalmaneser died and was succeeded by Sargon II of Assyria, who himself records the capture of the city. Thus, around 720 BCE, after around two centuries, the Kingdom of the ten tribes came to an end. The establishment of Judaism as religion was not complete until the Samarian Hebrew Pentateuch which had been written following the Babylonian destruction of the first temple, when temple priests retuned to Jerusalem with the original Hebrew Bible.

The establishment of religion of Judaism and the rebuilding of the Temple made the Temple priests the intermediaries if one wanted to communicate with God. This required Jews living outside of Jerusalem

---

[10] see 1 Kings 16:24

[11] see 2 Kings 17:5

to make pilgrimages to Jerusalem and offer sacrifices to the temple priests. The selection of the temple priests were strongly influenced by their benefactors, the Persian appointed governors who encouraged and provided financing for the rebuilding of the Temple. Following the Persian Empire allowing the Jews to return to Jerusalem, the Persian Empire crossed the Aegean Sea to conquer Athens. They succeeded but at a cost of resources that required them to withdraw. From this time forward the documentation of history increases substantially along with the development of Philosophy recorded in the Greek language. When Plato wrote his dialogue "The Republic" in 360 BCE, knowledge originating from Greek introspection of the human thought process and the understanding of the physical world, libraries of knowledge were established in greater numbers throughout the civilized world. Alexander had been tutored by Aristotle, the Greek language and Culture became dominant in all the conquered lands.

The consequence was that the culture of the communities away from Jerusalem became strongly influenced by Greek Culture often adopting the Greek language. Under this influence the "Hellenic Jews" differentiated themselves from the Jews of Jerusalem who adhered to the Traditions established by the Torah and the Hebrew language in which it was written. The Hellenized Jews began to disregard many of the Traditions that were established in the Torah, especially circumcision which the Hellenized Jews and Greek converts found repulsive when applied to adult males. This was especially true in the newly established cities of Alexandria and Antioch. The friction between the two groups resulted in a revolt against the Seleucid Rulers of a region of Judea that had achieved semi- autonomy from 140 BCE and independent from 110 BCE, until the kingdom was conquered by the Roman Republic in 63 BCE. The revolt became known as the Maccabeus Revolt and was not a revolt against authority as much as a revolt against Hellenized Jews. The revolt failed and a substantial population of Hellenized Jews grew in the cities of Antioch to the north and Alexandria to the south.

All Jews, Hellenized or not, had one common belief, God promised that if they followed his commandments, God would send them a Messiah that would restore their control over the promised land. The Messiah was not God, or a spiritual representative, but a Physical Human Being that would use God given powers to establish a Kingdom on earth that would be like heaven on earth.

Hellenized Jews established Synagogues not as places of worship, that could only be done at the Temple in Jerusalem, but as a community where the Hebrew Bible could be studied and the interpretation debated. There were people in the community who were not ethnically Jewish that found the belief in a monolithic God appealing and were welcomed to participate in the debates. There was one significant impediment to adult males converting to Judaism and that was the requirement of circumcision. Hellenized Jews began to consider circumcision and many dietary restrictions not part of Gods commandments, unnecessary and even barbaric. These differences between the orthodox Jews of Jerusalem and the Hellenized Jews was the situation when the Roman Empire acquired control of what was territory controlled by Alexander the Great.

When Alexander the Great died, the territories acquired were divided among his Generals. The Macedonian General Ptolemy consolidated all the territory of Egypt and established the Ptolemy dynasty. Seleucos, Alexander the Great's favorite companion, with the help of Ptolemy, conquers Babylon and founds the Seleucid dynasty in 312 BCE retaining control of the Levant until conquered by the Roman Republic in 63 BCE. The Seleucid Empire allowed the King of Jerusalem virtual independence who resisted the authority of the Roman Republic. The Roman senate appointed Herod who had converted to Judaism, King of Judea and provided him with the resources to lay siege on Jerusalem. In 37 BCE the last King of an independent Israel was captured and sent to Rome for execution. During this period of transition, Julius Caesar had changed the Roman Republic to a dictatorship that ended with his assassination in 44 BCE and began a period of struggle for control of the empire that was resolved when Augustus Caesar became Emperor in 27 BCE insuring that Rome had absolute control of the entire Levant. The Temple in Jerusalem continued to worship their monotheist God but the selection of Temple Priests needed approval of the Roman Authorities. Differences between the orthodox Jews of Jerusalem and the Hellenized Jews could only be resolved by a Messiah. Magi who were advisors to Kings, interpreted dreams and changes in the stars, predicted the coming of a Messiah.

Civilization of the world and the Cultural beliefs that produced those civilized societies had become documented in the written languages of Aramaic, Hebrew, Greek and Latin. The interpretation of these documents provides humans the ability to understand how evolution changes the future environment both physically and culturally, that can predict the future of mankind and the alternatives that are available.

The events that follow resulted in the evolution of two religions who's roots were in the original Hebrew Bible but redefined the meaning of Messiah. The first was Christianity and the Second was Islam.

CHAPTER XII

Evolution of Cultural Bias

Part 3 (Christianity & Islam)

The event that initiated the Common Era was the birth of an individual who developed a following as a Rabbi and was crucified on a cross sometime around 30 CE[12]. The Crucifixion became the event that launched the second major religious group referred to as Christians. This event was not considered significant for several years after the crucifixion of Jesus, by his brother James and a few apostles of Jesus.

Jesus was a Rabbi of Judaism who's followers proclaimed him to be the fulfillment of the Jewish prophecy that God would provide the believers of Judaism a Messiah. Following his crucifixion, Jesus' brother James, established a council of rabbis that promoted the belief of Jesus the Christ (King of the Jews). The council consisted of twelve dedicated followers to represent the twelve tribes of Israel some of which were Apostles of Jesus and some disciples who were dedicated to the message that Jesus taught. Information about the development of Christianity during the first generation following the Crucifixion of Jesus Christ comes almost exclusively from Saul of Tarsus.

Saul of Tarsus was a Jew who's parents were Jewish merchants that were enslaved when Rome took control of the Levant and taken to Tarsus in a province where their talents could be useful and their beliefs were not threatening. They became prominent citizens and were granted Roman Citizenship. Their son Saul, as a Roman citizen, was given an education that included learning Greek philosophy, and Latin law. As a Jew of the Benjamin tribe. Saul claimed to have underwent a thorough rabbinic training in the Pharisaic schools "as to the law of the Pharisee"[13]. As a educated merchant and Roman citizen, he was able to travel freely and spent significant time in Jerusalem were he became a

---

[12] CE is the now excepted designation for the Common Era replacing AD and it's religious connotation.

[13] see Philippians 3:5

guardian of the Temple, a position that enforced the authority of the Temple Priests who considered the followers of Jesus as a threat to their authority.

Three years after Crucifixion, Saul, who changed his name to Paul, made a trip to Damascus were he intended to enforce the Temple Priests assertion that Jesus was a false prophet. During this trip Paul had a revelation were he was informed by an angel sent by God, which informed him that Jesus was the Messiah not just for the ethnic Jews, but for all that believed in him as the Messiah. Paul returned to Jerusalem and meet with James and his "right-hand man" Cepheus, better known as Peter the Apostle, and informed them of his revelation. They accepted his revelation with the caveat that any gentile who accepted Jesus as the Messiah would have to convert to Judaism.

Nothing more was heard from Paul for 17 years. There are only passing remarks that Paul made in his letters to the Galatians. He tells us that he "conferred not with the flesh and blood but went away into Arabia for three years before retuning to Damascus".[14] After 17 years had elapsed, he visited Jerusalem were he again met with James and Peter. After a fortnight in Jerusalem, Paul took himself off again, this time to Syria and Cilicia.[15] By early 50 CE, Paul had arrived in Corinth then crossing the Aegean sea to Ephesus and ended his first mission as an Evangelical returning to Antioch, one of the cities established by Alexander the Great that had a significant population of Jews.

Populations of Jews distant from Jerusalem developed synagogues not as a place of prayer and worship, but rather as a place to discuss how they should practice the teachings of the Torah in their community. The discussions often included people of the community who were sympathetic to Judaism but were not Jews.

The Jews of Antioch had adopted the Greek language and became an ideal location for Paul to preach. Preaching to anyone who would listen and convert to become Jews, those who believed in Jesus as the Messiah, was his goal. Conversion to Judaism required the circumcision of all males which proved to be a great obstacle for adult male gentiles. Paul suggested that circumcision was not required for adults which prompted

---

[14] Galatians 1:13

[15] Paul: The mind of the Apostle, A.N. WILSON, page 81

James to send Peter from Jerusalem who informed Paul that such a compromise was not possible. The dispute caused Paul to leave Antioch and Peter becoming the leader of all the Synagogues in Antioch, a position that would later be designated as Bishop. After training another Antioch Rabbi as his replacement, Peter returned to Jerusalem. Paul continued his missionary journeys to Greek communities increasingly converting gentiles to Judaism that excepted Jesus as the Messiah and minimizing the need for circumcision or the need for strict adherence to dietary restrictions. It was Paul who introduced the tradition of Communion, symbolizing the eating Christ's flesh and drinking of his blood that further antagonized the Orthodox Jews.

Paul began his Evangelism in 51 CE in Antioch, the third largest City in the Roman Empire, after Rome and Alexandria. Paul taught that the return of Jesus was eminent in his lifetime and that the only requirement was to spreading the word across the Empire that believing in Jesus as the Messiah, would insure their participation in "Heaven on Earth". Paul clearly believed that Jesus' return was eminent but not until the people of the empire were given the opportunity to believe in Jesus the Messiah. His desire to visit Rome on the way to Spain in the context of "never to preach were Christ's name had already been heard"[16], was clearly stated in Romans 15:22-25 Preaching the word of Jesus in Spain, the most western edge of the Empire at the time, negated the need to establish a permanent Church or the Temple. The anger against Paul from the Temple Priests and Orthodox Jews threatened his life should he return to Jerusalem. However, Paul had promised James that he would collect a tithing to give to the poor, and on completion of his fourth missionary journey, he returned to Jerusalem with the tithing and was arrested. He would have been executed but evoked his right as a Roman Citizen to be tried by the Emperor in Rome. When Paul arrived in Rome, he was placed in House arrest but was allowed to communicate with those who wanted to learn more about the teachings of Jesus and continue to write to the other communities that he had visited. The last chapter of Romans consists of all the names significant in spreading the word of Jesus in the city of Rome, Peter was never mentioned.

---

[16] see Romans15:20-21

The first documented Christian Bishop of Rome was Clement I, how he was chosen is a matter of speculation. Centuries later Church Historians introduced the stories of Peter being the first Bishop of Rome and his Martyrdom in Rome. Evidence of Peter being in Rome was first introduced when the Cathedral of Saint Peter was built on the top of Vatican Hill and a Tomb was uncovered during the excavation for its foundation. The Tomb had an inscription on the side that could not be deciphered except for the word Peter. The assumption was that it was the tomb of the Apostle Peter. Vatican hill was part of a Pagan graveyard during the lifetime of Jesus and Vatican Hill was reserved for honored Pagan preachers. The title given to those so honored translated into the word peter. After Rome burned, Christians were crucified in large numbers and the pagan graveyard was opened to Christians but that would not account for Peter given an honorary gravesite. During the time of Jesus their were many preachers who claimed the ability to perform miracles. One in particular claimed to have the ability to perform all the miracles that were credited to Jesus. Simon was his name and he was baptized by Philip the Apostle[17]. The baptism took place in Samaria, following the baptism Simon preached to the Samaritans that his powers to perform miracles was God given and that qualified him to be the Messiah which resulted in a confrontation with Peter[18]. Simon is known to have gone to Rome where he continued to encourage the idea that he was the Messiah. He is far more likely to have been buried in a tomb on Vatican Hill than Peter. In any event Clement I was the head of the Christian church of Rome but encouraged the thought that he had no more authority over Christianity than any other Bishop. This was confirmed when the Emperor Constantine declared himself the secular head of the Christian Church but allowed a council of Bishops to arbitrate differences between them regarding theology. The first theological disagreement that required resolution was the spiritual identity of Jesus.

A monk in Alexandria preached that Jesus was a Human chosen by God just as Prophets were chosen even though his mission was to provide a path to salvation for all mankind. The Monk and Presbyter,

---

[17] see Acts 8:12-13

[18] see Acts 8:18-24

Arius, used the Gospel of John[19] as his proof where Jesus was quoted as having said "There is one in Heaven greater than I" This was in conflict to those who believed that Jesus was the incarnation of God as a Human Being. The Emperor called for a council of Bishops to resolve the mater. The Bishops meet in Nicaea and was the first Ecumenical council. After two months of intensive debate and pressure from the Emperor Constantine who favored Jesus as being the inspiration that lead him to become Emperor, the majority of the Bishops agreed on a creed, known thereafter as the Nicene creed. The Nicene creed included the word "homoousios" meaning consubstantiation or one in essence which laid the groundwork for the Trinity as the official church doctrine. The controversy continued which lead to several more councils. In 331 CE, six years after the council of Nicaea and a year after establishing the city of Constantinople as the new Capitol of the Empire, Constantine commissioned Eusebius of Caesarea to write 50 bibles in Greek for use by Bishops appointed by the Emperor. His son was Jerome who became was anointed a priest by the Bishop of Antioch and became a student of the Christian Bible, went to Rome for the Synod of 382 CE and so impressed Pope[20] Damasus I that he made Jerome a permanent member of his council. Jerome undertook revising previous versions of the bible based on Greek manuscripts of the new Testament which later became the primary source of the Latin Vulgate. The oldest know biblical text that included the 27 books of New Testament and almost certainly a source used by Jerome in writing his Latin translations, was a codex[21] that was written in a form that was adopted by the early Christians and have been found in Alexandria and Rome. There are four codex that contain the text of the Bible, the oldest written by the Bishop of Alexandria in 367 CE referred to as Codex Sinaiticus. An almost identical list of biblical text and possibly a copy with some text differences that did not include "Revelations" in the New Testament, can be found in the in the Vatican Library and named the Vaticanus codex. The caudexes were written in Koine Greek using script called uncial that included only capital letters. A third codex called the

---

[19] see John 14:28

[20] a title that was also used in reference of the Bishop of Alexandria a Antioch

[21] a predecessor to the book, later referred to as a manuscript (manually written).

Alexandrian codex where it originated was taken to Constantinople. The Gospels of Christ that were incorporated into the new testament were selected by a council of Bishops who decreed that those Gospels not accepted should be destroyed. Destruction of these documents, just as in Egypt, made determining the truth of the events in the new testament much more difficult.

The first Gospel that was written in the new testament was the Gospel according Mark. Mark was born in what is now Libya, but was then the western edge of Egypt and spoke the Coptic language. The Coptic language was the language spoken by the common people of Egypt after being conquered by Alexander the Great. It was a mixture of ancient Egyptian and Greek using the Greek alphabet in it's written form. When Mark's family moved from a rural community west of Alexander, they brought their son Mark as a young boy to Jerusalem where he could be educated in the teachings of Judaism. It was there that he learned of the teachings of Jesus and was befriended by the Apostle Peter who may have been related to Mark's parents. Peter encouraged Mark to become fluent in the Greek Language and after Peter's confrontation with Paul, Peter had Mark establish a church in Alexandria.

The Gospel of Jesus according to Mark was not Mark's account but no doubt Peter's account that he told to Mark. The Gospel ends with an appearances by Jesus where he declares "Go out to the whole world; proclaim the Good News to all creation. He who believes and is baptized will be saved; he who does not believe will be condemned."[22] This was a mandate for the establishment of a Christian church not intended for Paul, or James, which were no longer alive, but intended for Peter.

During this time, Rome had begun a siege of the Temple and would send all Jews, including Christian Jews, into exile. The Messiah was not God but one who was sent by God to overthrow those who ruled over the Jews and therefore, a constant threat to Roman authority. The Orthodox Jews were banned from Rome, but the Christians were allowed to stay because some Romans came into contact with Hellenized Jews and were converted. This changed when the city of Rome burned in 68 CE and the Emperor Nero blamed Christian Jews of starting the

---

[22] see Mark 16:16

fire. Soon afterward, the Orthodox Jews and Christian Jews combined forces one last time to overthrow Roman domination but was crushed in 70 CE. The Gospel according to Mark made clear that the separation of Christianity from Judaism was complete.

The story of Peter being rock upon which the Christian Church was built, was introduced after the Bishop of Rome declared himself to be the soul head of the Christian Church. The actual development of Christianity and it's leadership in Rome can be more accurately determined by studying the letters of Paul. There emerged four apostolic churches as Holy Sees of Christianity; Jerusalem established by James- Jesus' brother, Antioch established by Peter, Alexander established by Mark as a proxy of Peter, and Rome that through historical manipulation was credited to Peter. There is no verification that Peter was ever in Rome and was never mentioned by Paul during the time he was in Rome.

Mark was strongly influenced by both Paul and Peter having traveled with both, Paul arrived in Rome around 60 CE when he was placed under house arrest. During the time of house arrest, he was allowed to communicate with Roman Christians and write epistles to the Christian Jewish Communities he had established. There is no record of Paul standing trial or being executed, but by the time that Rome burned in 64 CE, Paul has disappeared. With Paul gone, the spreading the word of Jesus as the Messiah was left to James' council of Christian Jews. There was a great deal of turmoil in Jerusalem and James and his council were very unpopular with the Temple priests which lead to the James being martyred in 62 CE. The Gospel according to Mark was written sometime between the disappearance of Paul who believed that Jesus the Messiah would return in his lifetime. The establishment of a church was not necessary. With Paul not wanting a successor and James dead, the Apostle Peter would presumably be his successor. A new church would be required but could not be located in Jerusalem with the Roman authority determined to crush a Jewish uprising and the failure of the rebellion in Jerusalem, was that there was no Bishop who claimed to be the Spiritual head of Christianity.

The second Gospel to be written was the life of Jesus according to Mathew described those events preceding and immediately following the birth of Jesus. The story of King Herod and his encounter with the Magi is not mentioned in any other Gospel. There was a King Herod though he died in 4 BC according to the Gregorian Calendar, thus changing the year 0 as the beginning of the Common Era (CE) and changing the birth of Jesus to 6 BCE plus or minus two years. If there is any truth behind the story, Joseph and Mary fled to Egypt where Jesus spent his childhood. This can be confirmed by events that took place in Egypt, show the likelihood of the families being present in Egypt. The family returned to Nazareth in time for the events described by Mark to take place. Mathew also elaborated on the Crucifixion and the events following the Crucifixion adding the Transfiguration. The Transfiguration[23] defined Jesus as sole spokesman for God replacing the laws and rules defined by Moses and Elias. The revelation was made to Peter, James, and John (the Baptist?). This was an elaboration of the Transfiguration described in Mark's Gospel, but the only way that the story of the Transfiguration could be known to Mark is if it was told to him by a Eyewitness. That eyewitness was certainly Peter.

The next Gospel was the Gospel according to Luke written at least a decade after the Gospel according to Mathew and was an elaboration of the first two Gospels. Luke was also responsible for recording the Acts of the Apostles. The translation of the old testament combined with were the new testament became the Christian Bible that eventually became the Latin Vulgate. In 380 CE, Theodosius I, the Empire was permanently divided between the West, regions of the Empire, declared Christianity the Empires sole authorized religion. Within the next 100 years, the Byzantine separated from what was to eventually become the Holy Roman Empire.

A fourth Gospel of Jesus according to John, presumed by many to be John the Baptist and one of Jesus' closest Apostle, who is also thought to be the author of Revelations. The events described in Revelations did not describe the life of Jesus, but rather prophesized the spiritual consequences. Revelations was ambiguous as to when such events would take place, but implied certainty that the events would end on judgment day, the last day of Human life on the planet earth. There are very few Christian churches that do not include Revelations in the New Testament and was used by most churches to threaten their parishioners with the consequence of not accepting the churches authority.

---

[23] see Mathew 7-1-9

In 398 CE, a synod of Bishops was held at Hippo-Regius, North Africa[24], in order to canonize the New Testament as the official doctrine of Christianity. The Bishop of Hippo-Regius was Augustine whose mother was a Roman Christian but allowed Augustine to pursue the education of upper class Roman citizens. In 386 CE, under the intellectual influence of the Bishop of Milan, at the age of 31, Augustine was baptized a Christian and pursued his philosophical interests as a church theologian, soon to become the Bishop of Hippo-Regius. The Synod of 398 CE was the first of a series intended to establish a Cannon[25] for Christianity culminating in 419 CE with a Synod held in Carthage defined as "the code of canons of the African Church" written in Latin but reflected and quite possibly was the translations of the uncial Codex's described in the previous paragraph. The Latin Vulgate, originating from Jerome's translations of Codex Brixianus, an old Latin version, and corrected using the Alexandrian Codex, emerged during the same period and was eventually adopted as the Canon of the Roman Catholic Church. Regardless of which source was used in any particular version of the Bible, they all have essentially the same text that provides the forensic clues to determine the accuracy of the stories.

Even during his time, Theodosius I's Empire was in decay with the Barbarian invasions resulting in the fall of Rome and the end of the Western half of the Roman Empire in 476 CE. From this time forward the Bishop of Rome had to fend for himself as the Emperor of Constantinople was not willing or able to spend the resources necessary to defend the Bishop of Rome. When Barbarians conquered any city, they kill the leadership and the population that did not flee that would be a source of future rebellion including women and children. This did not happen to the Christians in Rome. It is presumed that the Bishop of Rome was able to convince the Barbarians that he would assure the Barbarians that the Christians would not rebel if they were spared. This saved his life as well as the Christians. It also promoted the belief that Rome was an Apostolic See.[26] This belief would require that the first Bishop of Rome, either was an Apostle or appointed by one. The

---

[24] now known as Annaba, Algeria

[25] authoritative scripture that define the beliefs of a religious group.

[26] Episcopal jurisdiction that attributed to the authorization by an Apostle.

Apostle that was promoted by church theologians at the time was Peter, James's presumed successor. The First Ecumenical Council convened by the Emperor Constantine in 325 CE, had the primary purpose of resolving Arian controversy as to the spiritual nature of Jesus that was resolved with the Creed of Nicaea, later modified and became known as the Apostle's Creed establishing that the "Son of God" was begotten by the Father and had no beginning (incarnation of God). It also established the first Canons of the Christian Church to be universally excepted in the Roman Empire. The sixth Canon declared Apostolic authority to See's[27] from which the Bishops of Alexander, Antioch, and Rome presided with special consideration for the Bishop of Jerusalem as the seventh Canon.

In 451 CE, The emperor Marcian convened the Council of Chalcedon, that issued the Chalcedonian Definition that declared Jesus to have had two natures one spiritual and one corporal, incorporated in a single entity, further enforcing the belief in the Trinity. The Coptic Christians of Egypt are one of the few Christian groups that did not and do not accept the Trinity. The council also added many cannons, the twenty-eighth cannon stated that See of Constantinople, the new Rome, was second in Honor and Authority only to Rome itself, putting it ahead of Alexandria and Antioch.

By command of the Emperor, the population of the empire had turned Christian and the kings had appointed bishops that were subject to the ecumenical law of the nearest Apostolic See. The Bishop of Rome used his influence to get the protection the Kings could offer, in return the Bishop designated their claim to be King as blessed by God. The need to show that Bishop of Rome was Apostolic grew with stories being introduced that Peter had appointed Clement I as the first Bishop followed by stories of Peter visiting Rome and being Martyred there.

The Emperor Justinian I, realizing that he had lost control of much of the western Empire, sought to re-conquer the lost territory and succeeded in restoring control over all territory along the coast lines of the Mediterranean and connected seas including those on the coastline of Italy, and all of Egypt plus the coast line cities from Egypt to the Atlantic ocean. A plague that lasted four years, prevented him from re-conquering Gaul and the territory north of the Danube. Justinian I

---

[27] Chair or Throne

introduced the Institutes of Justinian, a codification of Roman Law that became the textbook for law students, allowed judges to hear evidence deemed relevant resulting in decisions that were more objective and the bases of Common Law in many countries to this day.

The cost of re-acquiring lost territory, the plague, and internal corruption forced the Byzantine Emperors to tax all transactions that alienated both Christian and Jewish communities. This made the expansion of Islam over Byzantine territories popular among the Christian and Jewish Communities. The Muslims initially treated those communities fairly, taxing them the same as they taxed converts to Islam and allowed them to enforce their own laws according to their beliefs.

The story of Islam is the story of Mohammad who claims to be the last prophet. The book that defines the beliefs of Islam was dictated by the prophet himself and not a collection stories handed down over generations. This text called the Qur'an, was not just about a single man claiming to be the last prophet, but a translation of God's thoughts by the prophet. Not only did he author the Qur'an over his adult lifetime, but in the last half was in a position of power. No other prophet was in a position of power when making their prophecies, but rather challenged those in power of what would happen if they did not heed the prophets warnings. The influence of power can be seen if one compares the first half of the Qur'an the second half.

When Mohamed the prophet died, he was both the spiritual head of Islam and the head of Government. He left no successor to the throne and it was decided that his father-in-law, having married the Daughter, Aisha, and was Mohammad's close advisor, would be the first Caliph[28]. His successor was not a blood relative, but a disciple of Mohamed and was known for his pious nature. The third Caliph, Uthman, was a son-in-law of Mohamed and, his Army conquered North Africa from the Byzantine Empire, and raided Spain, conquering the coastal areas of the Iberian Peninsula, as well as the islands of Cyprus and Rhodes. Sicily was raided in 652 and the Sasanian[29] Empire

---

[28] one who is the political and religious successor to Mohamed

[29] The last vestige of the Persian Empire consisting of everything East if the Levant through what is now Iran.

was conquered extending Islamic Empire's Eastern frontier to the lower Indus river. Many Muslims, especially Egyptian Muslims, resented reforms initiated by Uthman that took away advantages that they had acquired over non-Muslims and newly converted Muslims. Uthman called for a assembly at Mecca where he promised that their concerns would be addressed. Muawlyah came to the assembly and saw that Uthman was in a dangerous environment. He suggested that Uthman come to Damascus with him. When Uthman refused to leave Medina, Muawlyah offered to send an army contingent to Medina to protect the Caliph. Uthman refused on the grounds that it might incite civil war. Less than a year later, in 556 CE, Ulthman was assassinated. Uthman's greatest accomplishment was the codification of the Qur'an using the Arabic alphabet. His successor was Ali, cousin and son-in-law of Mohamed and was the only person born in the Kaaba[30]. Ali's appointment as Uthman's successor was a compromise that was not unanimous. When Ali refused to go after those who assassinated Uthman, civil war broke out and Ali moved the Capitol from Medina to Kufa, a Muslim garrison in Iraq. Shia Islam arose from resentment of the Caliph Uthman. Shia Muslims believe that Mohamed designated Ali to be the first Imam[31] and the only one qualified to be a Caliph. Muawlyah who was related to the Caliph Uthman was appointed as the governor of Syria and the Levant. The taxation imposed on Jews and Christians that did not feel represented by the Emperor aided the rapid expansion and provided a base on the Mediterranean Sea where they built a navy. In 655 CE, the Muslim[32] Navy Defeated the Emperors Navy destroying 500 Byzantine ships.

Resentment of Muslims against the Shia claim that Ali was the only legitimate Caliph caused Ali to agree to a treaty following a uprising at Siffin[33] in 657. Even many Muslims who supported the Shia claim that Ali was the only Legitimate Caliph, did not support his reforms of taking

---

[30] building in the center of the most sacred Mosque of Mecca

[31] Shia muslins define Imam to be the leader of Islam and the only person qualified to be a Caliph.

[32] An Arabic word meaning one who submits to God, and follows the religion of Islam.

[33] Located on the banks of the Euphrates River in Syria.

from the rich and giving to the poor. The treaty was between Ali and Muawlyah resulting in the recognition of the Muslims that believed in the Umayyad school of Islam lead by Muawlyah. Muawlyah's army was then able to take control of Egypt and the Arabian peninsula. Egyptian Muslims, who were particular offended by the policies promoted by Ali, went to Kufa and assassinated him in 661. Muawlyah was appointed Caliph which began a Caliphate based on heredity that became known as the Umayyad Dynasty. Sunnis accept the first four Caliphs as the Rashidun Caliphs or Righteous Caliphs. The Sunnah[34], along with the Qur'an, established the canon of Sunni Islam. As in Christianity, this lead to conflict between the groups with one group defined as Sunni's and the other Shia. The Caliphate that controlled the most territory, was the Umayyad Caliphate established in Syria and included Syria, much of what is now is now Iraq and Iran, the Levant, the Arabian peninsula, and North Africa from Egypt to Iberia. The Shia school of Islam was reduced to parts of Iraq and much of the present day Iran, but accounts for no more than 10% of Muslims. The Umayyad Caliphate ended in 750 except for a caliphate established in Iberia that lasted until 1031 when the last Muslim stronghold was drive out the Iberian Peninsula.

The Umayyad Dynasty was replaced by the Abbasid Caliphate by appealing to non-Arab Muslims and claims of being directly related the uncle of Mohamed. In 750 the Umayyad Caliph was defeated and fled to Egypt where he was assassinated. Immediately after his victory the first Caliph of the Abbasid Dynasty, Abul Abbas As-Saffah, sent forces to Central Asia and established his Capitol in the new city of Bagdad where the worlds first Paper mill was built using Chinese slaves. Bagdad was established on the Tigris River in 762. As-Saffah's inclusion of non-Arabs in his Court and Army as well as the representation of Jews, Nestorian[35] Christians and Persians in his government, resulted in the translation of Greek philosophical and mathematical knowledge into Arabic. During the Abbasid Dynasty, the position of Caliph became

---

[34] Verbally transmitted teachings, deeds and sayings, of Mohamed plus reports about Mohamed's companions.

[35] Nestorian Christians believed that Jesus was both a Human and Spiritual Entity, a belief that was compatible with eastern religions that every individual has both identities.

largely ceremonial with the authority and leadership resigning in the Prince and Commander in Chief designated as the Emirate[36]. The head of government of several nations that have adopted Muslim Theocracy, are referred to as Emirates to this day. The Dynasty ended in 1220 when the Mongols captured Bagdad.

In 795, the Bishop of Rome, Pope Leo III was elected Pope, in haste on the very day that his predecessor was buried and immediately informed Charlemagne, who had consolidated central Europe, that Leo III had been unanimously elected to show that he regarded the Frankish King the protector of the Holy See. By request Charlemagne sent letters of congratulations and stated that it was his function to defend the Church and it was the Popes Function to pray for the realm and for the victory of his realm. In 799 Pope Leo III was attacked presumably by his predecessors relatives who attempted to cut out his tongue and gouge out his eyes, accusing him of Adultery and Perjury. He was rescued by king Charlemagne's solders and soon fled to King Charlemagne's camp and was received with great honor. Charlemagne had Leo III escorted back to Rome and in November 800 CE Charlemagne himself went to Rome and held Council on Dec. 1st. On Dec. 23rd, Leo III took a pledge of Purgation and his opponents were exiled. Two days later in Saint Peters Basilica on Christmas day, Pope Leo III Crowned Charlemagne as the Emperor of the Holy Roman Empire. Charlemagne was unaware of the plan, later said that he would not have attended the ceremony had he known. It was his decision to be protector of Church and not the prerogative of the Holy See. Charlemagne issued a directive requiring all his subjects to learn how to read and write and charged the Bishops and their Presbyters with the task. Due to hereditary rules, the Empire was divided among the male heirs, disputed territory resulting in two groups of kingdoms, one group that became France and one that were Germanic.

Otto I was able to unify the Germanic kingdoms and in 962 CE, was crowned by Pope John XII resurrecting the Holy Roman Empire. Otto I believed as Charlemagne, that his authority did not come from the Papacy and expanded the empire by defeating the Hungarian invasion and befriending the Byzantine Emperor to the concern of Rome. Otto I sent his army to take control of northern Italy, twice including

---

[36] Dynastic Islamic Monarch.

Rome when he appointed Leo VII. When Leo VII was deposed, he retook Rome appointing Leo VIII as Pope. The animosity between the aristocratic families of Rome and their ability to choose the next Pope as well as the Roman Pope's primacy over the Pope of Constantinople continued to grow. The Ecumenical Patriarch of Constantinople, Michael I Cerularious antagonized the Ecumenical Patriarch of Rome by writing a letter referring to him as brother rather than Father. This and other antagonistic acts against Latin Christians practicing their religion according to direction from the Holy See of Rome. In response, the Pope Leo IX sent the Patriarch of Constantinople a letter in 1054 CE, that cited a declaration of Constantine, a Decree that bestowed on the Pope of Rome supremacy over the other four principle Sees Alexandria, Antioch, Jerusalem, and Constantinople. The Decree has since been shown to be a forgery in 1439-1440 CE. Cardinal Hubert was sent as a Legatine having the authority to define and impose Church Doctrine. Cardinal Hubert delivered a notice of excommunication to the Patriarch on July 16, 1054, three months after Leo IX died. The Patriarch Michael in turn excommunicated the Cardinal and Pope of Rome. Regardless of the legality of the excommunications, their were now two Christian Churches that divided the Empire into the Latin speaking Kingdoms of the west with Rome as it's Holy See, and Greek speaking kingdoms with Constantinople as it's Holy See. Constantinople was also Capitol of what was now the Byzantine Empire. Rome was not the Capitol of the Holy Roman Empire who's Authority conflicted with the authority of the Holy See. That conflict could not be resolved so long as the Emperor had a say in the selection of Roman Pope.

All three Religions, Judaism, Christianity, and Islam, claim to know the intent of God and that includes the intent to degrade all other faiths. The more extreme interpretations also include the use of torture and murder. Examples abound in the confrontations between God and those adversaries as cited in the Torah. In the Qur'an, justification is provided by the Prophets own actions and stated belief in Qur'an that such actions are permissible if necessary to insure that Islam is accepted as the superior Religion. Christianity has demonstrated such behavior most notably during the Crusades and Inquisitions. These aberrations of God's intent were the result of social bias from the culture in which the founders of these Religions existed.

CHAPTER XIII

Evolution of Cultural Bias

Part 4 (Roman Catholic Church)

Control of Human Behavior is the Goal of any institution be it secular or religious in nature. Conflict between the Governmental institution and the most influential Religious institution for control of Human Behavior was inevitable. The schism that separated the Holy Sees of Rome and Constantinople in 1054 CE, did not resolve the conflict between the Holy See of Rome with the Holy Roman Emperor. The conflict could not be resolved so long as the Emperor had influence over the selection of the Pope. In 769 CE, a synod in Rome decided that the Pontiff would be elected by Cardinal[37] Deacons[38] and Cardinal Priests[39] subject to the approval of the Emperor. In 845 CE Cardinal Priests of Rome were selected to serve as Cardinal Legate[40]. In 1059 CE, Pope Nicholas II in the synod of Lateran, established the College of Cardinals for the purpose of eliminating any authority of the Emperor or Italian Aristocracy in choosing the Pope. The conflict between the Emperor and the elected Pope as to who had the authority to select the next Pope, continued 300 plus years, There was a constant conflict between Italian Aristocracy determining who would elect the next Pope; the College of Cardinals, or the Holy Roman Emperor. The conflict was more complex because the Pope needed the help of the Emperor to fend of invasion of a Byzantine supported group from southern Italy. The conflict continued through the crusades and the rise of the Ottoman Empire that threatened Christianity in Europe.

Normandy was a consequence of Vikings invading and establishing a settlement that eventually became a part of France. As Normans

---

[37] a title bestowed by Papal appointment.

[38] an ordained minister of an order below a Priest.

[39] Cardinal Priests are Bishops who serve in Diocese outside of Rome.

[40] Representative of the Pope empowered on matters of faith for settlement of ecclesiastical matters.

they then invaded southern Italy in conflict with forces supported by the Byzantine Emperor. Pope Benedict appealed to the Holy Roman Emperor, Henry II, to help the Normans and Papal troops to defeat the Byzantine supported forces. This was followed with Papal support of the Normandy invasion of England in 1066 that reinforced the authority of Holy See of Rome in England and the end of Paganism.

A symbiotic relationship had developed between the Kings of the Empire and their Bishops. The Bishops encouraged the collection of revenue from the parishioners of their diocese for the King who return rewarded the Bishops handsomely. A portion of the revenue collected was shared both with the Holy Roman Emperor, and the Holy See of Rome that resulted great concentration of wealth and the consequential power of these two entities to influence Human behavior. Their desire to increase their wealth and influence, as is true of all Empires, resulted in a desire to increase the size of the Empire. Their was little room to expand to the West and known wealth to be obtained in the expansion to the East. In 1095 CE, The Byzantine Emperor Alexios I Komnenos sent an ambassador asking for help against Muslim Turks who were threatening Constantinople. Letters written by Pope Urban II say that his goal was to wrest the Holy Land and the Eastern churches form the control of the Seljuk Turks[41] never mentioning Jerusalem specifically. He granted remission of all their sins to those undertaking the enterprise to liberate the eastern churches. It was the crusaders themselves who specifically identified Jerusalem as their target of liberation from the Muslims who conquered it during their rapid expansion. The Crusaders took Jerusalem in 1099 CE, killing every man women and child even though many of them were Christian. Pope Urban II died 14 days later and never gave a response to the slaughter.

The Temple Mount on which a Mosque was built, was given to a Christian group who's mission was to fight for the poor and protect Christian Pilgrims on their travels to the Holy Land. In 1339 CE, more power was conferred on the order by Pope Innocent II, who issued the Papal Bull[42] *Omne Datum Optimum* that stated that the Knights of

---

[41] a Sunni Muslim Dynasty

[42] A letter issued by the Pope of Rome with a seal (bulla) that authenticated the source.

the Templar could pass freely through any border, owed no taxes, and were subject to no one's authority except that of the Pope. Conquered territory, once pillaged of its wealth, became a liability that required investment in order to restore it's value. The Knights of the Templar were in a unique position to take advantage of their ability to make such investments. Officially the idea of lending money in return for interest was forbidden by the Church, but the Knights of the Templar sidestepped this with the stipulation that the Templar retain the rights to the production of mortgaged property meaning that instead of charging interest, they charged rent. Templar became a financial power and the majority of the Order's infrastructure was devoted not to combat, but to economic pursuits. Territory acquired by the Knights of the Templar, lead the Order to a position of significant power, both in Europe and the Holy Land. They owned large tracts of land in Europe and the Middle East, built Churches, and Castles, bought farm and Vineyards, were involved in manufacturing and import/export, had their own fleet of ships and for a time owned the entire island of Cypress.

In 1145, Pope Eugene III called for a second Crusade to the Holy Land in order reclaim territory acquired by the first crusaders. The second Crusade consisted of two armies, one lead by Lois VII of France, and the other by Conrad III of Germany. In order for these Kings to reach Anatolia and the Holy Land, they needed to cross Byzantine territory. Both Armies were defeated by the Seljuk Turks who received secret aid from the Byzantine Emperor Manuel I Komnenos. An ill-advised attack on Damascus ended the second Crusade that resulted in the eventual fall of Jerusalem in 1187 CE that gave cause for the third Crusade in 1189, against Saladin who was the first Sultan[43] of Egypt and Syria. The third Crusade ended with a treaty between Richard the Lionhearted and Saladin that left Muslim control of Jerusalem but granted Christian Pilgrims[44] and merchants access to the city leaving only three Crusader states; Tyre, Tripoli, and Antioch. Their was soon a call for a fourth Crusade to recapture Jerusalem this time by entering the Holy Land through Egypt. This would require the crusaders to go to North Africa by Sea and that would require building a large fleet and

---

[43] Sovereign of Muslim territory.

[44] people who travel to a sacred place as an act of religious devotion.

training a large number of men to sail the fleet across the Mediterranean Sea. Venice was the only port capable of such a task but would have to suspend most of its trading operations and that would demand a large fee. An agreement was reached that the Crusaders would pay the cost and Venice began the task of building the fleet. When the Crusaders arrived in Venice, they numbered less than twenty-five thousand, the Venetians had built a Fleet and trained crews able to carry thirty-five thousand Crusaders as agreed. The Crusaders were unable to pay their fee and a compromise was reached that would require the Crusaders to pay their fee by intimidating ports and towns in the Adriatic Sea ending in the sack of Zara on the coast of Croatia. These were ports that the Byzantine Emperor Justinian I had re-conquered. The pope of Rome was explicit that no Christians, Byzantine or Roman, should be hurt by the Crusaders and later issued a warning that any Crusaders who did so would be excommunicated. The leaders of the Crusaders suppressed the information in their desperate attempt to acquire the treasure necessary to pay the Venetians. After sacking Zara in Nov. of 1202 CE they decided to wait for winter to end before taking the next step.

In Constantinople, a coup removed the Emperor Isaac II Angelos from the Throne, blinded him, presumably so that he could not rule the Empire, and put him in prison replacing him with Alexios Angelos, the Emperor's older brother. The son of Isaac II approached the Crusaders and Venetians in Zara, and offered to pay the Crusaders debt to the Venetians and escort their fleet to Egypt if they would depose Isaac's brother. The Crusaders and Venetians agreed, but the son of Isaac realized that he could not pay the Venetians, decided to attack the Venetian Fleet by setting his won fleet on fire and Ramming the Venetian ships. This strategy failed and the Crusaders sacked Constantinople in July of 1203 CE, pillaging and destroying many Christian relics. When Innocent III heard of the conduct of his pilgrims he was filled with shame and rage, and he strongly rebuked them but still excepted the treasure that was brought back. The Latin Church maintained an uneasy control over the Byzantine Empire for the next 57 years before outside forces made Latin Christians control of Constantinople impossible.

The ability to develop banking practices by the Templar's was greatly enhanced by the publication of the book *"Liber Abacci"*[9] in 1202 CE, that introduced the so-called *modus Indorem (method of the Indians)*, today known as Arabic numerals. The book advocated numeration using the digits 0-9 and place value. The book showed the practical importance of the new numeral system by applying it to commercial bookkeeping, conversion of weights and measures, the calculation of interest, money changing and other applications. The book was immediately well received throughout Europe and had a profound impact on European thought.

The Knights of the Templar no longer had a presence in the Holy Land and the King of France had inherited an impoverished kingdom from his father who was deeply in debt to the Knights of the Templar. King Philip IV imposed a tax on all religious organizations which resulted in conflict between the King and the Pope Boniface VIII. King Philip sent a party to kidnap the Pope, who refused to resign so they beat him and left him for dead, supporters rescued him and brought back to Saint Peters Cathedral, where he died in three days having a fever and strange malady that expressed itself as a state of madness. The Collage of Cardinals selected a Pope who would not be hostile to Philip IV, and who was quick to release King Philip from excommunication put upon him by Pope Boniface VIII. He also annulled his predecessor's Bull *Unam Santam,* that asserted papal supremacy secular rulers and Papal authority was derived from the Apostle Peter. Pope Benedict XI was Pope only eight months when he died suddenly suspiciously in July 1304 CE. The election of the new Pope had to be held in Perugia where Pope Benedict XI had died. The conclave that elected the next Pope were equally divided between Italian and French Bishops and lasted nearly a year. Finally Pope Clement V was neither Italian or Cardinal, was elected in June 1305 CE probably as the result of backroom politicking by King Philip. On election he was urged to return to Rome but Pope Clement V chose Lyon, France for his Coronation. Pope Clement V brought Francesco Petrach to Avignon as his chief advisor. Francesco's father was an aristocrat who provided his son with the best education available. Francesco was sent to Athens to study Greek philosophy, then to Montpelier, France and Bologna, Italy; the world's oldest universities to study Law. He developed a passion for language and became a poet-deplomate for several Popes. He defined the "Dark Ages" as the time period between the fall of Rome and the Renaissance. Francesco recommended that Pope Clement V reverse Papal supremacy over secular Rulers which lead to the arrest of hundreds of the Knights of the Templar in France on Friday the 13th, October 1307 who were charged with heresy among other charges. A council refused to find the Templar's guilty of heresy but Pope Clement V abolished the order. King Philip expropriated the Templar's bank. In 1309 CE the entire papal court moved to a province located in France but owned by the Holy See of Rome. The city in which the Pope took residence was Avignon. Residence of the Papal state paid no taxes and was a haven for French Jews who were poorly treated in France proper.

The death of Clement V in 1314 CE, was followed by an interregnum[45] of two years. In 1316 CE a Papal conclave of twenty-three Cardinals was arranged in Lyon, France, elected John XXII who established his residence in Avignon, continuing the Avignon Papacy. Pope John XXII opposed Louis IV of Bavaria which prompted Louis IV to invade Italy and in Jan. 1328 had himself crowned the Holy Roman Emperor by the Roman Senate. Three months later he installed a Spiritual Franciscan who John the XXII hated, Nicholas V. Nicholas V was excommunicated and later on assurance of a Pardon made a confession of his sins to Pope John XXII who absolved him. When Pope John XXII died, in 1342 CE, a Papal conclave, on the first ballot, voted Benedict XII[46] Pope. The custom at the time was to vote for an improbable cardinal on the first vote to see how other Cardinals were voting but the strategy resulted in an unintended Cardinal being elected. Pope Benedict XII reversed the policies of Pope John XXII, made peace with Lois IV, and came to terms with the Franciscans. He promulgated an Apostolic Constitution *Benedictus Deus* that defined the Church's belief that the souls of the departed go to their eternal reward immediately after death as opposed to waiting for Judgment day.

Pope Gregory XI returned the Papacy to Rome in Jan. 1377 in an attempt to resolve schisms within the Church. He died the following year, in Mar. 1378. The following month a papal conclave consisting mostly of French Cardinals, under pressure of a mob demanding a Roman Pope, elected a person who was neither French or Roman, and not a Cardinal, Pope Urban VI. The Animosity against the French Cardinals was so great that they reconvened the Papal conclave electing Clement VII who took residence in Avignon and though not appointed

---

[45] A period when normal government is suspended between successive reigns.

[46] Roman Catholic Holy See recognizes him as Benedict XI

by an Emperor, was later considered to be an Antipope. The immediate problem was to resolve what authority was greater than the authority of the Pope and could resolve the conflict. A reform movement that held that the authority of the Church resided in the Ecumenical Council and not the Papacy, (referred to as Councilarism), inspired a council to be reconvened in order to resolve the schism. A church council was held at Pisa, Italy in 1409 under the auspices of the Cardinals to try and resolve the dispute, but it added to the problem by electing another incumbent, Alexander V. He reigned briefly to his death when he was succeeded by John XXIII who won some but not universal support. Finally, a council was convened by Pope John XXIII in 1414 at Constance, Germany, to resolve the issue.

The council was the largest and culturally significant gatherings in the history of medieval Europe. The council lasted for more than three years. 700 high ranking Church officials from all over Europe, and the Christian Byzantium Empire, as well as almost every city an feudal state in Europe, Middle East, North Africa including representatives of the major universities, schools, and Humanists[47] attended.

Humanism emphasized the value of Human Beings, individually and collectively, that prefers individual thought and evidence[48] over established doctrine or faith (fideism). Francesco Petrach was encouraged to become a priest where he would be put on the fast track to becoming a Bishop because of his knowledge in theology an philosophy. He refused to take the vows required to become a priest which would restrict his ability to travel as a diplomat and pursue his intellectual adventures. During his travels he studied the writings of Cicero, Augustine, and had in depth discussions with Thomas Aquinas that resulted in him publishing *Secretum meum*; an analysis of those discussions. It was in that analysis that he stated that human nature allows the individual to recognize and pursue good without God's intervention and therefore saving one's soul is possible without the grace of God or even being aware of God. Such observations resulted in his becoming known as the father of Humanism. *Secretum meum,* he points out that secular

---

[47] A Renaissance cultural movement that turned away from medieval scholasticism and revived interest in ancient Greek and Roman thought.

[48] Rationalism and Empiricism- knowledge comes from sensory experience.

achievements did not necessarily preclude an authentic relationship with God. He argued instead that God had given Humans their vast intellectual and creative potential to be used to their fullest[11].

The Council of Constance rapidly evolved into an unofficial international book fair in which the Humanist perspective made a significant contribution. Much of what follows in this chapter can be found in the book *The Fourth part of the world*[10], along with more detailed supporting information. Geographical texts figured prominently including a new translation of *Claudius Ptolemy's Geography* which introduced the concept of identifying locations by Latitude and Longitude. This represented the first exchange of knowledge that gave the educated people of Europe the benefit of knowledge acquired by Greek Christians since the Schism between Latin and Greek Christians. The council finally resolved the multiple Pope issue by electing Pope Martin V in Nov. 1417.

Once Portugal realized that the world was a globe, they believed that if they could sail around Africa, of unknown size, they could reach India and Indonesia without crossing hostile territory and possibly a much shorter journey. Since North Africa was controlled by Muslim States, the Portuguese Navy launched a surprise attack and captured the fortress of Ceuta that overlooked the Straights of Gibraltar. They gained access to West Africa which allowed the King of Spain, with aid from the Pope, to expel the Moors[49]. In 1455, the Pope Nicolas V issued a Bull *Romanus Pontifex,* that granted Portugal the excusive rights to Ceuta and discovered lands of West Africa.

A map had been produced that showed Asia and what is now known as Japan, to be relatively short distances from the north end of West Africa. This was based on a miscalculation on the circumference of the World. A letter from the Author was sent to Christopher Columbus who had appealed to the king of Spain for sponsorship of a small fleet to sail west from West Africa in hopes of reaching the west coast of India. After three years of consideration, and with no other direction for Spain to increase it's territory, Queen Isabel agreed to sponsor the fleet of three ships. The first of four trips that Christopher Columbus made in 1492, resulted in the discovery of the Bahama Islands but what Christopher thought were islands off the coast of India or Asia. On his

---

[49] African Muslim people of mixed Berber and Arab descent.

return, the Pope granted all rights to all discoveries west of a line of Longitude that would include the discovered islands and everything west. This line was designated the Line of Demarcation in the treaty of Tordesillas between Portugal and Spain. On the third trip, Christopher reached what we now know to be Central America with a Navigator by the name of Amerigo Vespucci. When Christopher returned without bringing back any significant gold or silver, he lost favor the King and Queen of Spain, who then offered to license other Captains wishing to explore the new world. The business partner of Amerigo Vespucci. Was the first to request and receive a license. When Amerigo Vespucci reached the new world at a point several degrees south of the equator, Amerigo sailed south along the coast of what is now known as South America which caused him to turn east. Although he claimed that strong currents caused him to turn back, he no doubt realized that he had crossed the Line of Demarcation into Portuguese territory. When he returned to Spain and reported his discoveries, he resigned his license to sail for Spain. Amerigo then convinced the King of Portugal to sail with a reconnaissance mission to the new world were he followed his previous path across the Line of Demarcation and established what is now Brazil as Portuguese territory. When Amerigo returned to Lisbon, he felt that he was not properly accredited for his contribution and returned to Seville, Spain were he acquired a position in the *Casa de Cotractacion -* House of Trade, a government agency established to oversee trade with Spain's overseas possessions.

With so much new information accumulating about the *new world,* there was significant demand for a map of the world that included this new information. There were two distinct skills required in the development of such a map. One was that of a cartographer who could give an orientation of the shape and relative location of all landmasses. The other was to locate and label all locations of significant Human habitation. Because of Amerigo's influence, the new southern continent became known as America and the portion claimed by Portugal was named Brazil.

The cultural framework that was forced on the indigenous people of the new continent was imposed by a Roman Catholic Church that had become the single most powerful entity in the world. The northern continent was not explored and exploited until another century had passed, primarily by the English and the French. Although the English had a strong Christian bias, the Reformation marginalized the influence of the Roman Catholic Church. Even so, many of the cultural biases of the Protestants and the Roman Catholics were the same.

The success of the Reformation was the result of a single technological advance, the Printing Press invented by Johannes Gutenberg in 1440 by combining movable type with the screw press allowing 3600 copies of a single page to be printed in a single workday. This allowed people who had been taught to read[50] to acquire knowledge that was otherwise only acquired by lectures from the pulpit that was totally controlled by the church. The book that ignited the Reformation was a translation of the Ninety-Five Thesis written by Martin Luther, that criticized the Roman Catholic Church for their dispensing Indulgences[51], in Jan. 1518. Martin Luther's popularity grew rapidly because of Catholic's rebellion against the corruption implicit in the actions of the Roman Curia[52]. Publication of a translation of the Vulgate[53] into common languages so that people could read the teachings of Jesus Christ and pray to God directly, met with fierce opposition from the Catholic Church taking the form of the Inquisition. The dispute between the King of England and the Pope resulted in the establishment of the Church of England in 1534 that defined the initial cultural influence in North America as Protestant[54], mostly Anglican[55]. Non-monolithic structure of the Protestant movement, allows for changes in cultural bias, though difficult, more easily achieved.

---

[50] Originally by order from the Emperor Charlemagne.

[51] Remission of part or all of the temporal and especially purgatorial punishment that according to Roman Catholicism is due for sins whose eternal punishment has been remitted and whose guilt has been pardoned

[52] Administrative arm of the Holy See.

[53] Latin copy of the Bible that became canonized by the Roman Catholic Church.

[54] Member of any non-orthodox Christian Church that separated from the Roman Catholic Church.

[55] Considers the Holy See of Canterbury replacing the Holy of Rome.

The most fundamental, and difficult to change, cultural bias that applies to nearly all religions of the world, is gender bias. The reason for this bias is not only that the stronger side will always dominate, but that the male provides the seed and that the female has been assumed, until relatively recently, to be the incubator of the seed. Even today, many of those in religious leadership positions, refuse to except the biological equality in determining the traits of the offspring. Because this bias was written into religious scripture that has been designated as *"Gods Word"*, this bias will continue until the majority except that religious scripture is mythology containing both wisdom and falsehoods. Men and women have different attributes that justify discrimination, These differences giving women access to healthcare not required by men and to facilities needed for personal and maternal hygiene.

The second fundamental cultural bias that also applies to nearly all religions of the world, is that God picks winners and losers, rewarding the winners and punishing the losers. The justification for the bias is again rooted in religious scripture that cites disasters as acts of God as punishment for Human behavior. Again overcoming this bias will require the majority to except religious scripture as being mythology.

There is a third cultural bias that is more insidious than the first two that justifies behavior justified by the cultural indoctrination of morality. All religions have been guilty of such indoctrinations, but the Roman Catholic Church has made behavior changing indoctrination a trademark of it's rise in power beginning with the Crusades through the Inquisition and manipulation of Governments throughout the world, setting an example for other religions of the world.

<u>Willful Ignorance</u> is the consequence of cultural bias. The indoctrination that promotes cultural bias insures that prejudice results in denial of any contradicting information and promotes hatred toward anyone who promotes an alternative view. Those that control wealth often promote prejudice in order to gain support for their own objectives. Their objectives often are contrary to the needs of those who are co-opted into supporting those objectives. Only through the education that promotes the critical thought process that recognizes personal prejudice, can such prejudices be overcome. The critical thought process is the true source of free-will.

CHAPTER XIV

Legacy and the Future

Life is a series of experiences, the experiences available to an individual is limited both genetically, and by the environment in which one must exist. Good and Bad things happen, not as a consequence of any Super-Natural intervention but rather as a consequence of the complex evolutionary interactions between the laws of nature, even if the evolutionary process was the result of a Super-Natural event. This evolutionary process has resulted in life on the planet earth culminating in the evolution of mankind, a process very probably duplicated throughout the universe. Science has provided sufficient data that shows that a source of energy exists that can only be described as Super-Natural. How this energy can be accessed in a way that will benefit mankind, is the challenge that baffled mankind from the beginnings of civilization. Many believe that it can be accessed through meditation including prayer, but this method of access has not yet been confirmable. Mythology has obscured the search for evidence of Super-Natural Energy, but new approaches in the search promises to shed new light on the nature and source of Super-Natural Energy and how it gave rise to sources of energy with which we are familiar.

Through the evolution of Homo-sapient, Humans have acquired beliefs handed down to them through generations of cultural indoctrination based on stories that could not be confirmed. As time passed, these stories changed from presumed factual descriptions of what happened to mythologies that included exaggerations and improvised embellishments. Life is the ability to react to the environment. Human life adds the ability to control one's life through the exercise of Free-Will not requiring praying for intervention of a super-natural entity.

Over the past few centuries, Mankind has acquired the ability to change the course of Evolution not just by significantly changing the environment but by changing the DNA of all life forms and their ability to respond to the changing environment.

All philosophical discussions, be they discussions with respect to religion, politics, economics, or any subcategory, can be reduced to discussions regarding Human Behavior. What you believe certainly effects your behavior, but what you do, determines who you are. Too often one's actions do not reflect the fundamental beliefs of that individual, but rather attempts to achieve personal gain. What the individual does or does not do, can be determined by measuring the motivation on a scale that ranges from fear to desire. The emotions that determine the position on the scale can be altered through the use of reason. Reason does not necessarily result in a better course of action. If the information used in the application of reason is misleading or completely inaccurate, the result will be misinformed decisions. Cultural bias is the cause of accepting misleading and inaccurate information especially if acceptance promises to provide personal gain.

Every government on the planet earth <u>should</u> have a constitution that states the goal of providing equal opportunity for all it's citizens fundamental human rights and provide for the education that promotes this goal. The legislating body establishes the laws that regulate Human Behavior in pursuit of these goals and provide for economic stability with equitable distribution of wealth. The economic achievements of any government can be measured as the Gross Domestic Product (GDP), which is the total domestic expenditures for all goods and services consumed domestically, corrected for those goods and services exported and imported. The domestic source of that expenditure is the consumer, be that an individual, a privately controlled entity, or the government. Any reduction in spending from any of these **will** result in job loss that must be compensated by the other sources of expenditure if one wishes to avoid a decrease in GDP. All sources can borrow in order to maintain or increase their level of spending, but such debt must be paid back with interest over the log run. How much debt can be carried without the excessive risk of bankruptcy, is a debate that will continue so long as there are lenders and borrowers, but spending must be maintained or the economy **will** shrink. A high per-capita GDP indicates a strong economy but does not say anything about distribution of that wealth.

The roots of modern humanism began with the Renaissance that as defined by Francesco Petrach who was highly influenced by Cicero's *De Senectute* (On Old Age), he was moved to kiss the book and thought that Cicero should have been beautified as Saint Cicero. Conflicts between belief in a supernatural entity as defined in religious theology and the ability to lead a good life without the aid of such theology continues to the present day and has been analyzed in depth by A.C. Grayling in his book *The God Argument*.[12] This argument suggests that religious theology and humanism are fundamentally incompatible and that secular Humanism is the only form of Humanism that meets the precepts of modern Humanism as defined by Corliss Lamont in his book *The Philosophy of Humanism*.[13] These precepts do not deny the existence of God as an entity but only that such a God does not change the physical environment for the benefit or detriment of any individual. The core of any constitution should include these precepts in establishing individual Human Rights. Dignity of the individual is the central Humanist value requiring protecting civil liberties that allows for freedom to express one's opinion, a right to non-violent protest, freedom of association with any group that does not promote violence as a method of change, separation of government regulation and the government imposed cultural bias that conflicts with individual Human rights, political democracy that does not restrict any citizen access to the ballot box, and most importantly a fair and independent judicial process. These Human Rights goals can only be achieved by insuring that all citizens be provided access to education that allows the citizens to maximize their potential to contribute to society, and access to food and healthcare for all citizens.

All governments regulate the economy which has two components. The first is a component that benefit's the society as a whole and insures the equitable distribution of wealth to it's citizenship, referred to as Socialism. The second component benefits those who produce goods and services to meet the demand of the consumers, be that individual, government entities, or privately owned entities. The cost of production requires the accumulation of capitol, resulting in the second component to be referred to as Capitalism. Capitalism requires the concentration of wealth to a minority group of individuals. Growth requires both components, even though there is a constant struggle for each component to dominate the other. A system of government that insures equal representation and regulation of both components requires a form of government that can transition from one group of leaders to another without revolution and the violence that is implied. A Constitutional Democracy guarantees such a transition, a Republic allows for the functionality of representative Democracy. Equal opportunity for all it's citizens to participate in both components of the economy and have an informed vote on their representatives is the only assurance that cultural bias will not prevent the long term success of a Constitutional Democratic Republic.

In the final analysis, the legacy of any individual is not just bases on one's accomplishments, but more significantly on the effects of one's accomplishments on the lives of others. What one accomplishments depends on two factors, opportunity and the ability to recognize and take advantages of those opportunities. Those opportunities are far more available in a society that adhere to the principles of a Constitutional Democratic Republic. The ability to take advantage of those opportunities is both genetic and the acquired education, formal and informal, that one acquires. All actions involve risk, the possibility of failure, but failure does not mean disgrace. Not taking action on an opportunity that would be a benefit to you and others, is a lost opportunity.

Whatever you decide to do within the framework available, will effect not only your future, but the future of everyone and everything that is effected by your behavior. Education insures that one will be able to make informed decisions that will have the greatest for both the individual and society. The benefit on society depends on the values that have been engrained into an individual by the cultural bias of that society. These values are often based on blind faith introduced through religious indoctrination. Part of every individual's education should be the rational evaluation of the major religions, particularly those most influential in one's social culture.

Altruism is the extreme example of a positive legacy that makes fame and fortune difficult. Both fame and fortune can be achieved by mitigating the self sacrifice of the altruistic equation and allows one to choose a partner in life that shares those values. Some self sacrifice is needed in the pursuit of any goal, if for no other reason than to appreciate the feeling of accomplishment. Pay attention to the needs of the next generation, they are the one's who will amplify your legacy and allow for the greatest appreciation of life. The goal in life should be clear, the challenge is to allow every Human Being the opportunity to achieve their goal. Establishment of governments that allow for the accumulation of wealth without the use of slave labor or the disfranchisement of those with minimal abilities to make a contribution.

Within the next few decades, there will be huge growth in the awareness of how the brain functions. This will result in understanding of those values that bring about change in Human Behavior. An article in the Sep. 20, 2013 issue of Science magazine refers to an on-line E-book by Tania Singer and Mathias Bolz entitled *Compassion Bridging Practice and Science,* that differentiates empathy and compassion, demonstrating how compassion for humanity can be enhanced through meditation. The ability of the majority to acquire awareness and control of compassion through education based on the principles of science rather than faith, will ultimately determine the future of mankind. This should be the goal of every individual, regardless of the strategy chosen for reaching that goal.

CHAPTER XV

Theory of Everything

Although evolution is continuous, the rate of change is not. During the evolution of the universe and the evolution of life on earth, there were relatively rapid evolutionary changes separated by periods of little change. The rate of change is a variable with a high standard deviation from the average over long periods of time. Data accumulated using the scientific method, shows that evolution continuous and the consequence of evolution always results in a more complex environment. The following summary of key events shows how, intended or not, the universe and life on the planet earth came to be as it is, and the possible future of mankind.

Approximately fourteen billion years ago, there was but a single force of nature in the universe - Gravity. The source of that gravity was matter, measured in units of mass. Matter has but one property, matter attracts matter. In order to do so, matter occupies space. The distribution of matter indicates that there is a source of energy that is opposite to the force that attracts matter to matter. Nether of these forces are detectable by the senses and are therefore referred to as Dark Matter and Dark Energy. When or how these forces were created is unknown, little more than speculation is the current status of these forces evolution.

Time is the measurement of change, if nothing happened, the their was no change and no universe. Searching for the truth is searching for what happened and when. We can only be aware of what happened through our senses. Electromagnetic radiation is the only force field that we are capable of sensing. The source of Electromagnetic radiation is Charge. What created Charge is also unknown but we have confirmable data that has identified the time that Charge came into existence at 13.8 billion years ago. Besides being a force-field, charge also has a property known as spin. Interaction between Charge and Dark Matter resulted in charged particles that formed atoms, initially Hydrogen atoms. The appearance of a neutrally charged particle called a neutron allowed atoms to form having nucleuses with multiple protons, when subjected to gravitational force that resulted in stars. Stars not only produced 100 plus different atoms called elements, that interacted to form molecules, but released enough Electro-Magnetic Radiation to elevate the temperature of surrounding planets, allowing for the evolution of life, at least on the planet Earth. Evolution of life began as single cell organisms over three billion years to evolve into multi-cell organisms that could more quickly adapt to the environment. Multi-cell evolution produced Human Beings capable of understanding the concept of Truth. The evolution of multi-cell life to Human Beings took over 630 million years, the evolution of Human beings to those being able to question the truth of their own existence has taken 10.000 years and continues to the present.

Truth is what happened, facts are what we believe happened. If you believe that a person told a lie, what the person said was the truth, your interpretation of what the person said was, based on your knowledge of what the person said, factually a lie. What you believe to be a fact may be correct but the truth is simply what the person said. What the bible said is the truth, your interpretation of what was said may be factual but not the truth. The more you become aware of what happened, the more you become aware of the truth. Your interpretation of what happened can only be truthful if your interpretation is totally objective. Acquisition of the knowledge required for an objective interpretation consumes energy and the resources needed to provide that energy.

The last four hundred years, marked by awareness of the world as having extensive but limited resources. Thus began the evolution driven by the desire to acquire control of as much of these limited resources as possible by the nations of the world. Much of this population growth was the result industrialization and technological advances that allowed for the exploitation of resources not previously available. The governments that encouraged these advances were most successful in acquiring the most resources.

Economics is the study of how resources are exploited and distributed. Exploitation of resources is most efficiently accomplished through Capitalism which results in the concentration of wealth. Equitable distribution of wealth can only be accomplished through Socialism. The government that provide the regulation that balances the two will be most successful in achieving growth while avoiding rebellion that results in inequitable distribution of wealth. A Constitutional Democratic

Republic provides the framework that allows such a government to exist. For the Democratic Republic to provide for the orderly transition of government leadership, all the citizens capable of making informed decisions must have access to the ballot box. The only qualifications required is that the individual is a member of the society that is in a government controlled territory, and that the individual is capable of making an informed decision (educated). Government regulation should have the intent that these requirements are meet and that no law is passed that would be act as a barrier to anyone who meet those qualifications, from voting. On the contrary, laws should be passed that encourage, or even require such people to vote.

In order that the United States Constitution more closely establish these principles, I recommend the following Amendments;

Informed Voter Amendment: In order for voters to have an informed opinion, individuals shall have the freedom to express their opinions on their beliefs and to disclose falsehoods regarding the beliefs of others. Philosophical, including religious beliefs or ethnicity, shall not be considered in determining congressional districts. Voters shall have access to the ballot box at least seven days before a federal election.

Equal Representation Amendment: Representatives shall be apportioned to the several states according to their population, as a percentage of the total United States population, rounded to a whole number, with all states having at least one representative. All states having more than one representative shall divide the state into contiguous districts having one major population center in each district and all districts shall have nearly equal populations. An Algorithm shall be provided by each state that determines the dividing line between the congressional districts. An additional representative shall be added for the District of Columbia and each separate territory not within a state.

Senate Representation Amendment: All States shall be ranked according to population. Those ranked in the lower third shall have one senator, those in the middle third shall have two senators, and those in the upper third shall have three senators. If more than one senator, the second will be elected two years after the first, and the third, two years after the second.

Term Limits: No government office, elected or appointed shall be held by a single individual for more than 24 years. Senate approval for any government position shall be voted on within 90 days of a nomination or the nominee shall be appointed to the position.

These Amendments would go a long way in insuring that the will of the majority is reflected the elections. Other Amendments should be addressed such as eliminating the Electoral Collage for electing the president, defining Corporations as economic entities not having the rights of individual and insuring that all economic activity be reported on an annual bases, including non-profit and charity organizations.

The future can only be predicted with a limited degree of reliability, but the size of life forms and resources available limit's their growth, requiring a relatively large amount of energy consumption per individual. This means that population growth will have to be limited if not reduced. Such population restrictions will be more difficult as the average life span increases.

The Supreme Court ruled that money promotes free speech and cannot be restricted by campaign law. Furthermore the source of that money need not be disclosed. The consequence has been that special interests have the ability to indoctrinate the voters with misleading and false information that disregards the democratic process. Legislatures, in order to attract the most money to their campaigns, become representatives of these special interests and not the voter. Proof of their influence is most apparent when over 90% of eligible voters want gun control legislation but the National Rifleman Association is able to control enough legislatures to prevent consideration of such legislation. The Supreme Court could reverse it's previous decision but an amendment to the constitution is the only permanent solution.

Imposition of successful population control will require change in the cultural DNA. The world population growth that will soon exceed the ability to provide for the distribution of food necessary to prevent starvation and the wars that would follow began around 1600. The population of the world in 1600, is estimated to be 500 to 580 million, by 1700, the population is estimated to have increased by 20%. From 1700 to 1800 an increase of an additional 30%, and from 1800 to 1900, an additional 43%. The increase from 1900 to 2000 was over 74% and the rate of increase is projected to continue to rise.

The Economic evolution and the evolution of Religious beliefs, are the two greatest cultural evolutions of the last four centuries and has resulted in a changed environment. As in all environmental changes, DNA either changes or goes extinct. So it will be with mankind. The changes will, with little doubt, be in the Brain rather then in physical attributes. Only through mankind's ability to respond to the changes in the environment, will survival of mankind and all other species be possible. A person of the next millennium will be fundamentally different then a person now living, even though they appear to be genetically related. Mankind's influence on the direction of evolution is real.

Mankind's influence on evolution will change the environment in a manner that will require the exploitation of resources necessary to insure species survival even as the DNA of the species evolves into a new species. Through science and technology, all threats to the species will be overcome. Death will become a choice and reproduction will be controlled. Programmed robotics will perform all tasks requiring labor. The goal of the new species that replaces Human Beings, will be to achieve a greater understanding of the interaction between the physical and supernatural life.

As the evolution of Human Beings continues, the environmental changes taking place today suggests that even within the next several decades, the changes will bring about another mass extinction. Mankind will most probably survive, but the new species that evolves will be one that rewards both empathy and compassion toward life as the greatest achievement possible for both the individual and society without the aid of an Super-Natural Entity.

Signs that the impending mass extinction is closer than most current predictions, are not only clearly indicated in the changes taking place to the Global temperature trends, but also in the crises defined by the inability to feed the rising global population. This crises has been clearly defined in the book *Grain of Truth by Stephen Yafa*.[14] In 1798 Thomas Malthus predicted that "premature death must in some shape or other visit the Human Race" with famine being the most likely cause. In the 1958-1961 time period, an estimated forty-five million Chinese died in the worst of all twentieth century famines. Because of its abundance and reliability as a source of protein and calories more than any other single food including rice, wheat has become known as the last defense against mass starvation. Western Economists predicted that mass starvation was the inevitable future following the population explosion after WW II. Norman Borlaug received a PhD in plant pathology and genetics from the University of Minnesota in 1942. He choose to pursue a career that would allow him to make a difference rather than accept a high paying position at large agriculture corporation that would have given him a laboratory for developing more profitable seeds. Instead he accepted a grant from the Rockefeller Foundation for doing research into the cause and remedy of a wheat disease known as rust that creating havoc especially in Mexico. He became the world authority on wheat by the time mass starvation was predicted to be eminent in the late nineteen-fifties. His efforts to develop wheat with short stalks that grew faster and produce significantly greater yields saved many lives in Mexico and later India, provided the tedious technique for finding a seed that was resistant to rust. Dwarf wheat that is rust resistant now accounts for 95% of all wheat production in the world. No wonder that he is credited with saving more lives than the efforts of all the religions combined and avoiding the beginning of a mass extinction. For his efforts he has been awarded the Nobel Peace Prize, the Presidential Medal of Freedom, and the Congressional Gold Medal.

APPENDIX

Mathematical comparison of Gravitational and Electro-Magnetic Energy

The following definition is from Wikipedia.org
Newton's law of Universal Gravitation: Every point mass attracts every other point mass by a force pointing along the line intersecting both point. The force is proportional to the product of the two masses and inversely proportional to the square of the distance between them. (Separately it was shown that large spherically symmetrical masses attract and are attracted as if all their mass was concentrated at their centers.) This is a general physical law derived from empirical observations by what Isaac Newton called Induction.

$$F = \frac{G(m_1 \times m_2)}{r^2}$$

Where:
- F is the force between the masses
- G is the gravitational constant
- m_1 is the first mass
- m_2 is the second mass
- r is the distance between the centers of the masses

The Electro-Magnetic force is proportional to Electro-Magnetic force and differs only in strength as can be seen by the equation that defines Electro Magnetic force.

$$F = \frac{K (q_1 \times q_2)}{r^2}$$

Where: F = force between charges
K = Electromagnetic Constant
q_1 = the first Charge
q_2 = the second Charge
r = the distance between the Charges

Although these forces are proportional, the Electro-Magnetic force field is 10^{36} times stronger in strength. The full strength of the Electro-Magnetic force is only felt over relatively short distances is because as soon as the force field leaves the Atom, there are competing Electro-Magnetic force fields that will tend to neutralize the accumulative effect except in special circumstances such as the atmospheric conditions of a thunderstorm or in power plants where the Electromagnetic force is deliberately amplified. The only force stronger than the Electro-Magnetic force is the Nuclear force field which is more than a thousand times greater than the Electro-Magnetic force field.

EPITAPH

To those whose feelings I have hurt, I sincerely apologize. To those who feel better for having known me, I only ask that you do the same for others. To those who have hurt my feelings, I hold no grudge but only ask than you refrain from hurting the feelings of others. Finally, for those who have made me feel better about myself, I truly thank you for the most precious gift a person can bestow on others. Writing this book has truly been an enlightening experience for me and hopefully for those who read this book with an open mind. Anyone who would like to ask a question or make a comment with the author may contact me via E-mail: tmoreau@juno.com. My Face book name is Thomas Moreau.

BIBLIOGRAPHY

1. University of Wisconsin-Madison, June 14, 2008, Planetary Science Letters: Ancient Mineral Shows Early Earth Climate Tough on Continents.
2. Cobb, Cathy - The Joy of Chemistry, 2005
3. Wired for Culture; Origins of the Human Social Mind. W.W. Norton & Company, 2012 ISBN 978-0-393-06587-9,
4. Vidal, Gore, Creation
5. Wilson, Edward O. - The Social Conquest of Earth, 2012
6. Mitchell, Stephen, - The Epoch of Gilgamesh, Free Press, Division of Simon & Shuster, 2004
7. Osman, Ahmed - Stranger in the valley of the kings, Harper & Row Publishers, San Francisco
8. Osman, Ahmed - Moses and Akhenaten, Bear & Company, Rochester, Vermont 2002
9. Sigler, Lawrence- Fibonacci's Liber Abacci, ISBN 0-387-95419-8
10. Famous First Fact International page 303, H.W. Wilson Company, NY 2000,
11. Lester, Toby - The Fourth Part of the World, Simon and Shuster Free Press, 2009
12. A.C. Grayling The GOD Argument, Bloomsbury USA, New York- Fist US Edition 2013\
13. Corliss Lamont - The Philosophy of Humanism, The Contimuum Publishing Co. Seventh Edition 1990
14. Stephen Yafa - Grain of Truth, Avery-Imprint of Penguin Random House, New York Copyright 2015

www.ingramcontent.com/pod-product-compliance
Lightning Source LLC
Chambersburg PA
CBHW030814180526
45163CB00003B/1280